FORKS, PHONOGRAPHS, AND HOT AIR BALLOONS

FORKS, PHONOGRAPHS, AND HOT AIR BALLOONS

A Field Guide to Inventive Thinking

Robert J. Weber

New York   Oxford
OXFORD UNIVERSITY PRESS
1992

Oxford University Press

Oxford   New York   Toronto
Delhi   Bombay   Calcutta   Madras   Karachi
Kuala Lumpur   Singapore   Hong Kong   Tokyo
Nairobi   Dar es Salaam   Cape Town
Melbourne   Auckland   Madrid

and associated companies in
Berlin   Ibadan

Library of Congress Cataloging-in-Publication Data
Weber, Robert J. (Robert John), 1936–
Forks, phonographs, and hot air balloons : a field guide to
inventive thinking / by Robert J. Weber.
p. cm.   Includes bibliographical references and index.
ISBN 0-19-506402-X
1. Inventions.   2. Creative thinking.   I. Title.
T212.W4   1992
609—dc20   92-22785

1 3 5 7 9 8 6 4 2

Printed in the United States of America
on acid-free paper

TO
Gloria, Mark, Karen, Clarence, and Edith
for their support and their ideas

Preface

I believe there are important general principles underlying invention, and our understanding of these principles can make us more appreciative of the built world about us.

Here, then, are my objectives. I want to place invention in the context it deserves. As the connection between technology and the creative spark or crawl of the mind, invention leaves clearer tracks across time and space than any other form of human creativity.

In the Prologue, I try to illustrate the richness and the penetration of the past in the simplest of present-day tools and activities. The earliest known inventions, in the form of stone artifacts, go back approximately two million years. From that time until the present there is a gradually accruing record of invention, and indirectly of the hidden principles and tactics behind those inventions. We will try to understand and appreciate the simplicity and hidden intelligence that underlie some of history's great inventions.

In Part I, I try to provide a context for invention. Chapter 1 offers numerous examples of invention's pervasiveness, along with a preliminary definition. Truly, we live in an invented world. Behind that world, it is argued, are many powerful principles that fuel and generate invention. A preliminary excursion into that world reveals the wonders of a coffeepot. Indeed, artifacts of all kinds—the results of the anonymous work of many hands and the hidden intelligence of many minds—yield principles that can be incorporated into the individual mind. Tentative distinctions are drawn between invention and related endeavors, such as design, applied science, art, and creativity. In Chapter 2, the inventions of children are examined and found to be rich sources of hidden principles. Novice invention is further examined in a class of college students asked to keep an invention diary to record the "bugs of everyday life." In Chapter 3, expert invention is studied by tracing a simple principle, the turn of the screw, through the inventions of Archimedes, Leonardo, and the Wright brothers. Abstraction and generalization distinguish the work of the experts from that of amateurs.

Part II introduces a framework for understanding invention. In Chapter 4 a descriptive scheme is developed, one with general categories like purpose, principle, and parts. This leads us into Chapter 5, where we begin to evaluate and compare inventions. Then in Chapter 6 we put the framework together to test our understanding of simple artifacts, like doorknobs and soup spoons, in the created world.

Part III introduces the idea of a heuristic which in this context is an idea generator. Chapter 7 argues that heuristics can serve as the engine of variation in producing invention, and Chapter 8 introduces some simple but important heuristics. With Chapters 9 and 10, we turn to more complicated heuristics—those based on the relations of multiple inventions. Chapter 11 shows how heuristics involving after-the-fact transformations can be used to organize and classify hand tools and architectural forms. Chapter 12 describes how to discover additional heuristics in the contexts of reinventing the wheel and sewing. And then Chapter 13 puts the section together with the argument that some inventions are after their time. Their late coming, it is argued, is due to the unappreciated importance of heuristic principles.

Part IV treats a number of common invention themes. For the most part, the principles already developed are drawn on to discuss these themes. Chapter 14 examines the materials that form the foundation of any invention. The chapter moves from the Plains Indians' use of buffalo to modern technology's diamond coatings. Chapter 15 develops the important idea of an interface, where two systems come together, and ranges from horses and carts to keyboards and people. Chapter 16 is a meditation on the idea of containment, a powerful notion that connects concrete cooking pots to abstract imprisonment. Chapter 17 explicitly introduces the important idea of procedural invention. To illustrate it, visits are made to the supermarket (where products are distributed) and the assembly line (where products are made). The chapter concludes with the procedure of maintenance (how to extend the life of products). In Chapter 18 the idea of transgenic forms is traced from its mythic origins on cave walls to the biotechnology lab. Along the way, a complex alchemy of procedures, materials, and existing inventions converge to produce new lifelike forms.

The Epilogue then tries to put everything together by looking at a setting to teach children about invention and to see the vistas of the invented world behind us and ahead of us.

Thus, this book is about the ideas that generate invention—the front end of invention. But as important as ideas are, they form but a part of invention's way. Concentrating on inventive ideas misses both the fine-grained detail often required for implementation and the broader canvas of social effects and reactions. The fine grain of implementation is often critical; putting the right nut and bolt together demands specialized knowledge and skill. I do not write about these nuts and bolts, nor the specific skills germane to a content area, because I am not sure that anything general can be said.

Similarly, I do not write of the history or social context of invention: the benefits and costs, how an invention will be received and used, and how a given invention is always embedded in a larger social system. These are all important concerns and they have been discussed ably by others.

Instead, this book is the result of steering a course between detailed technological knowledge and skill, on the one hand, and the social and historical context of invention, on the other. At this level I want to consider the conceptual basis of invention, where ideas come from and how they relate to one another. I want to convince you that much can be said about generating inventive ideas, that hidden intelligence abounds in the most common artifact, and that great generality is embodied in that intelligence.

My intention then is not to address the book principally to the aspiring or the professional inventor, but to all the rest of us who constantly invent new, useful, and nonobvious tools and ways to go about our work and leisure in the course of our everyday lives. Few of these inventions would be patentable, because others before us have solved the same problem or because the problem is not one with broad market appeal. These new inventions combine with past inventions. Moving in our everyday world is possible because inventions and their hidden intelligence convey us each step of the way.

There is much to be said for the wave of anticipation and insight surrounding the birth of an idea. I hope to contribute to our understanding of this experience

Albuquerque, N.M.   R. J. W.
August 1992

Acknowledgments

A book is the result of moving between the computations of solitude and the stimulation of others. I am grateful to many people for supplying that stimulation: Maria Ast, Stacey Dixon, Margaret Ewing, Sidney Ewing, Vicki Green, Becky Johnson, Antolin Llorente, Mike Lyons, Tim McCollum, Earl Mitchell, Carol Moder, John Mowen, David Perkins, John Solie, Jo Solet, and Gloria Valencia-Weber. All these individuals have influenced me, but as usual they should remain blameless for my excesses.

I want to offer special thanks to my editor, Joan Bossert, for her patience in teaching the art of book writing to a first-time author. The other members of the Oxford staff have also contributed greatly to my efforts.

Two reviewers also deserve much thanks. One, Mike Posner, has provided a helping hand and a critical eye for a number of my endeavors over the years; and Dean Simonton has offered many constructive suggestions. As usual, they are blameless for my excesses.

Contents

FORKS, PHONOGRAPHS, AND HOT AIR BALLOONS

Prologue

Peeling an Apple

I am sitting in my kitchen peeling an apple. In my right hand I hold a paring knife and in my left hand the apple. The knife has a blade and a handle. The blade has two working parts, the long sharp edge we slice with and the sharp point. The earliest knifelike form I know of goes back about two million years, to Africa. It is a rounded flat stone with an edge chipped sharp. No point is present; that was a long time in coming. The handle of my paring knife is wood, and it is attached to an extension of the blade. The idea of an explicit knife handle is probably not more than a few hundred-thousand years old; a separate handle other than stone is much newer. My knife's handle is fastened to the blade with rivets. The fastening idea is a profound one and probably began with lashing or tying things together. Fasteners make possible compound inventions composed of simple parts. The blade of my knife is no longer stone. It is steel, a material that first made its appearance a few thousand years ago. These are just some of the stories held in my right hand.

In my left hand, I'm holding the apple. It is a domesticated variety, an apple partly invented and shaped by the hands of many generations of growers, over a period of several thousand years. That shaping has worked to refine the apple's size, flavor, and color. My apple may even have come from a grafted tree, something that is truly the work of an inventive mind. These are stories held in my left hand.

My two hands now move together. I peel the apple by starting at

3

the top. I cut into it with the knife blade pointed toward me, thumb opposed to blade, and simultaneously rotate the apple in my other hand. A long spiraling peel descends because I try not to lift the blade. Why do I peel the apple like this? It can be peeled other ways.

A child peeling an apple does not have an adult's coordinated actions between left and right hands. She cuts a small piece of peel, lifts the blade, rotates the apple, cuts another small piece, and so on. The child's method, a piece at a time, is manufacturing by a series of discrete processes. The method I use—and that of most adults who produce a long uninterrupted peeling—is manufacturing by a continuous process. Continuous-process manufacturing is a more efficient method, with less lost motion from switching between actions like cutting and lifting. Discrete-process manufacturing is common in handcraft industries, and continuous-process manufacturing reaches its culmination in the modern assembly line of this century.

This story of the apple, the knife, and the two hands working together provides a first glimpse into our invented world.

I

THE HIDDEN
INTELLIGENCE
OF INVENTION

1

A Context
For Invention

I'm looking at a computer screen while my fingers are busily tapping at a keyboard made of a plastic. The entire computer is an artifact rather than a natural form. To be sure, natural materials are used to make the plastic of the keyboard and case, the chips inside the computer, the metal wires that connect the keyboard and the computer, and the glasslike face of the computer screen. But these natural materials are then processed into forms that do not occur in nature. Nor are particular natural materials essential to the computer as artifact. Almost certainly the present materials will be replaced in a few years with other materials, such as fiber optic cable instead of metal wire and a different plastic for the case. Ultimately, the computer is an artifact or invention of the mind. It consists of many fabricated materials that are new under the sun (and will be replaced by other new materials) and a concept. The concept is much more enduring than the physical components; it includes the ability to store and manipulate symbols, a perceptual interface to the human mind through a display (now visual output from the computer screen), and an interface for human motor activity (now a keyboard input).

As I look away from the computer, I see a wall covered with synthetic paint (probably a latex with artificial pigments) and a window sash of aluminum (a form of metal that must be significantly altered from its natural ore before use); a radio plays in the background; an air conditioner pumps out cool air; and an overhead light with an incandescent bulb illuminates my desk.

7

All of these things are human artifacts or inventions, external to ourselves. They are not part of the natural world of earth, plant, and animal. In fact, our lives are so shaped by the inventions of the mind that it is difficult to think of living without them, although people certainly have. That artifacts are so transparent to our eyes and other senses is a tribute to how closely a good invention fits between the changing world around us and the needs of our bodies and minds. A successful invention serves as an interface between environment and human need. That sounds good, but it must be qualified. The *need* spoken of may be apparent or it may be created by the attraction of the invention. No one needed a phonograph before Edison's invention of it in 1877. But after the invention people began to need it. In the most radical sense, we not only invent artifacts; we invent needs for them as well. Sometimes invention is the mother of necessity.

But what is the nature of that interface between our needs and the environment? Our clothes serve as an example. They modulate the effect of the external temperature on our bodies and offer a protective shield between ourselves and our environment. The reversible fasteners on these clothes—buttons, ties, zippers, Velcro—allow us to further adjust and fine-tune the microclimate about our bodies. These clothes and fasteners provide protection and comfort, and they allow us to inhabit environments unsuitable for naked bodies. Indeed, a very large part of the invented world is just such an interface: sometimes protective, sometimes comforting, and at other times enhancing or amplifying our limited senses, muscles, and memories. One more thing about clothes as inventions. The shirt on my back is 35 percent cotton and 65 percent poly-something, a suitable metaphor for the balance between the natural and invented worlds we live in.

But the inventions about us and on us do not end with our skin. Proceeding inward, I have fillings in my teeth made of alloys that do not exist in nature. I also have some surgical stitches in me that alter the natural configuration of my insides. Finally, my thoughts are shaped by symbols and concepts that are the hard-won mental tools of many generations.

Granted that invention pervades both our external and internal worlds, can we say anything general about it? After all, science has made momentous strides in understanding the world by using general methods and finding general truths. Can we also find general ways of going about invention? Are there general truths about the invention process behind my computer screen and embedded in the act of peeling an apple?

The introduction of a subject customarily begins with a definition, even if the definition is a fuzzy one. Certainly, invention is not easy to define, but we can begin with the criteria that the U.S. Patent and Trademark Office uses. To be patentable, an invention must satisfy the simultaneous criteria of being *new and useful and nonobvious*. The content categories of patentable inventions themselves have historically included *objects of manufacture or machines or processes or human-made materials*. In addition, the patenting of *asexual plants* was allowed in 1930, the patenting of *microorganisms* became possible in 1980, and the patenting of *higher life forms* (a mouse with a human breast cancer gene) became possible in 1988. The concept of patentability is evolving.

That brings us to the next point. While the Patent Office definition of invention is a good starting place, I want us to see invention much more broadly. Most of what we take for granted in life—from the clothes we wear, to the walls that we dwell within, to the invisible bonds that both join and separate us through credit and debt files and through supermarket checkout lines—is not patentable, but was once new, useful, and nonobvious to minds older than our own, minds that operated outside the context of any patent system.

Indeed, the workings of the mind and the ideas it generates are the ultimate source of invention. So viewed, invention is a potential bridge or connection between the worlds of technology and the cognitive sciences. While much of the creative, building side of technology and invention proceeds at an intuitive level, most of cognitive psychology, for all its methodological sophistication, has not dealt with the important problem of analyzing and systematizing the invention process. If we can find a way of connecting the constructive aspect of the technological and the analytical aspect of the cognitive, we may aid the development of both: The invention process becomes part of the larger context of human creativity, and the cognitive sciences are enriched by new and undeniably important problems.

Any rich approach to invention must apply both to the workings of the individual mind, the standard approach in cognitive psychology, and across minds, in which the collective efforts of many minds converge on the development of a particular tool or artifact. The first point is readily understood; we want to know what is going on in the individual's mind as he or she spins out inventive ideas, and we will make some guesses here. The second point requires clarification; it is in the tradition of the nineteenth-century psychologist Wilhelm Wundt who talked about a group or folk mind.

Indeed, most significant inventions are the products of many minds, often connecting the changes in a given artifact across vast expanses of time and geography. Inventions that are products of aggregate minds still may be analyzed for general principles of formation and change. Once principles are found, they may be incorporated by the individual mind.

This play between the individual and the aggregate mind is time honored and takes place in the formation of any complex artifact, from the formation of language, one of Wundt's examples and surely more than an individual effort, to the construction of a knife, a tool with beginnings in the Stone Age that is still evolving, to the design and construction of computers. Some cognitive scientists will find this approach uncomfortable, because a central dogma of the discipline is that the unit of analysis must be the individual mind. I disagree; such a narrow view makes impossible the study of invention, the best and oldest database we have for the understanding of human creativity.

The Invented World: Pathology or Hidden Intelligence?

I'm looking at a strange coffeepot design. It is Jacque Carelman's "Coffeepot for Masochists" (Figure 1.1). I saw it for the first time in Donald Norman's *The Psychology of Everyday Things* where it serves as an imaginative and comical case of bad design: to use it is to be hurt by it. Norman employs Carelman's coffeepot as a springboard for discussing bad design, ranging from VCRs that we cannot figure out how to program, to fancy telephones and computers that most of us do not begin to comprehend. Certainly, the examination of pathological design is worthwhile; we want the artifacts around us to be more responsive. But my approach will be just the opposite; I want to look at the hidden intelligence of everyday things.

To take up Wundt's point, many of the artifacts around us are in fact examples of hidden intelligences at work over hundreds if not thousands of years. Closer examination of Carelman's coffeepot reveals a design that is really quite intelligent. Suppose we introduce some variations into the coffeepot to help us see that intelligence. These variations constitute a *test for design*—they help us see what is functional and what is arbitrary, or perhaps what is an aesthetic as opposed to functional concern. Clearly, the handle is wrong; it belongs on the opposite side and up higher, where it normally is. In fact, putting the handle almost anywhere other than its normal position will adversely affect the functioning of the coffeepot, either by balance or by pouring coffee on yourself, as in Carelman's version. The usual placement of the handle is an example of hidden intelligence at work; and it is an anonymous intelligence, because the countless inventors or designers behind it are not known and probably can never be known.

We continue applying the test for design. Is the lid necessary? Take it off and see what happens; the coffee cools much more rapidly. The lid is there for intelligent reasons. On top of the lid is a knob. Perhaps an ornament? No, it helps us put on and take off the lid; it enables us to hold the lid and steady the pot while we pour; and its shape makes it easy to grab from any direction, with either hand. The narrow neck attaching the knob to the lid also provides some resistance to heat conduction, and it may keep the knob cooler than would a wider neck.

Figure 1.1   Carelman's "Coffee Pot for
Masochists." A bad design for the spout and
handle arrangement, but a good design for the
other features. Norman sees folly; I see hidden
intelligence; both views are correct. (Carelman,
1984; by permission.)

Evidently, the knob is a kind of handle, and its placement in the middle
of the lid is not an accident—anywhere else and it is unbalanced. Nor
is the length of the spout arbitary. If it were much shorter (say, one
inch long), it would be hard to pour and see a target cup at the same
time because the pot would obscure our view. If it were much longer
(say, fifteen inches), accuracy would be affected again, and the pot
would take up needless space while being less stable. The angle of the
spout's lip also reflects intelligence; it must be just right to prevent
dripping while pouring. The shape of the pot is important; the diameter is larger at the base than at the top. Why? The answer is forthcoming as soon as we reverse this configuration, with the small diameter
at the bottom and the large at the top. The usual arrangement ensures
stability.

   The coffeepot contains another and even more deeply hidden act
of intelligence. To see it, consider a primordial coffeepot or teapot. Likely,

it was a vessel with a single opening at the top. That opening probably served dual functions; it was the orifice both for filling and for pouring liquids. No doubt such a pot dripped during pouring, and, if it had a lid, that lid would have to be lifted each time we poured. A reasonable next step in the series of transformations was to bend the opening on one side into a lip specialized for pouring, a lip like those we see on water pitchers. While this solved the drip problem, it made fitting a lid much more difficult. The next step is to create a second opening. Once we imagine two openings, one for filling and one for pouring, we can independently specialize each one for its particular functions. The spout can be fine-tuned to just the right length and angle to ensure dripless action, and it can be made quite narrow to prevent heat from escaping. Once again, the top can have a lid to hold in the heat, and the lid can have a knoblike handle that is easy to grab from any direction, either to lift or to help steady the pot during pouring.

Now let's step back and see what has happened here. Two functions, filling and pouring, were originally yoked to the same spatial location, the single opening at the top of the primordial pot. Once these functions were recognized as being different, it became possible to separate them spatially. Once spatially separated, they could then be independently adjusted and fine-tuned for their particular function. This spatial separation of functions is a standard tactic in invention. Other examples range from simple ones like having separate In and Out doors to a restaurant kitchen, to complex ones like the assembly line in which different assembly acts are spatially separated and specialized. Who would have thought there is a deep similarity between a coffeepot, restaurant doors, and an assembly line?

These are just some examples of the hidden intelligence residing in the coffeepot. Like the act of peeling the apple, that intelligence is largely anonymous and cumulative, the product of many minds at work over a long time. As Wundt maintains, this is typically the case for artifacts that have been with us for some time. Just a little scrutiny will reveal their secret intelligence, especially if we apply the test for design by changing and rearranging features to see what happens. This *what-if play* provided by the test for design helps us find hidden intelligence.

Having provided a taste of invention, I want now to distinguish it from its near relatives. I make these distinctions with some trepidation; almost certainly it is possible to find exceptions to any list of contrasts. Nonetheless, I want to draw distinctions because distinctions are the raw material of thought and action.

First, consider the relative called *design*. Design is what many engineers consider themselves doing. The distinction between design and invention is not a sharp one. In the limiting case, invention is the creation of new forms, while design is the adapting of existing forms to present constraints. We design a house or a car to fit present environments and constraints, such as cost and the availability of materials, but at one time both the house and the car were inventions. So we can think of invention and design as on a continuum, from new form to familiar form with some new constraints. Much of the time, what creative people do is somewhere between the two. Buckminster Fuller's geodesic dome is part design and part invention. It is `old form (dome), old function (dwelling), all combined with a strikingly original mode of construction (the geodesic principle). So while the distinctions between invention and design are not easy to draw and often blur into one another, the emphasis of each is quite different, and I believe that different principles are frequently involved.

Next, I claim that invention is not merely *applied science*. Invention is much older than science; stone tools and other artifacts bear ready witness to this. Moreover, invention is universal while science is not. Every culture has its technology and invention, but the same is by no means true of science, a way of thinking that began with the Greeks.

An important difference between invention and *basic science* consists of the search path to the final result. In a common form of invention, one starts with observable and manipulable elements and combines these in different ways to produce complexity. A complex electronic device like a computer may contain millions of transistor-equivalent units in its hundreds of integrated circuits. But in basic science, one often starts with complexity and tries to find simpler hidden causes behind that complexity. A disease is tracked from its symptoms and distribution to its ultimate lair as a heritable defect, an errant gene on a particular chromosome. One way of capturing these two directions in the search path is to say that invention starts with the visible (the single transistor or integrated circuit) and makes it invisible (an unseen part in the finished computer). However, science works to make the invisible (the hidden causes, laws, and mechanisms) visible in the form of explicit laws and models. Said in another way, the product of invention is usually a tangible and working form, often with hidden mechanism. The product of science is much more abstract than that of invention; that product is most readily captured as strings of words, specialized symbols, and equations. Another difference is in evaluation. For invention the evaluative criterion or standard is workability or efficiency. But for science it is the understanding of a reality beyond the senses.

Now in what sense is invention related to *problem solving?* The work of Herbert Simon and his colleages has shown the importance of problem solving to higher forms of thought. The schematic structure of a problem is a goal for us to reach, a set of initial conditions to start from, and a set of operations to get from the givens to the goal. In many problem-solving situations, like games of chess, the goal, the givens, and the permissible operations are well known; the difficulty resides in stringing them all together in the right order. In invention often the goal is imprecise, and neither the givens nor the permissible operations are well specified. Recent work in the problem-solving tradition comes closer to invention; it is the study and recreation of scientific discovery through computer simulation. But even here the differences are still noteworthy. Interesting science is likely to be quite complex, so that a computer simulation may well be in order. Significant invention can be very simple in comparison. In fact, so simple that one can readily see what is happening without any aid from simulation. These are the inventions that will be of primary interest to us—those that are almost transparent. While in some sense invention is obviously a kind of problem solving, we will find many invention principles that seem somewhat different from those normally extracted in problem-solving studies.

What is invention's relation to *art?* Like invention, art is often a building-block activity in which simple elements or forms are combined to create complexity. The product of art is also a sensible form that again may conceal important structures and elements. Unlike invention, which is concerned with a workable product, art need have little or no emphasis on workability and efficiency. The evaluative standard for art is aesthetic and dramatic impact, matters of secondary interest to invention.

Lastly, consider invention's relation to *creativity.* I have suffered more than a little trying to make this distinction clear. Taking a composite of several dictionaries, we see that *to invent* is "to originate as a product of one's own contrivance, to produce or create with one's imagination." Notice how invention is defined in terms of creativity. The corresponding definition for *to create* is "to bring into being; cause to exist; to produce; to evolve from one's own thought or imagination; to be the first to represent; to make by investing with new character or functions." I don't see any sharp division here. The differences between the two definitions could probably be interchanged and no one would know it. According to these definitions, invention and creativity are synonyms, especially if we consider invention in a broader context than technology.

While I shift my positions on this matter almost daily, I presently see invention as a subspecies of creativity, one in which the evaluative standard is primarily workability—as contrasted, say, with the aesthetic standards of art.

With these distinctions between invention and its relatives in hand, we can state in a lighthearted way the two fundamental problems of invention. The first problem is not having enough ideas so you don't know what to do next. The second is having too many ideas so you don't know what to do next. Both problems are solved by having one good idea. Our task is to develop methods for finding that idea.

2

Novice Invention and a Problem-Based Diary

The curtains in my living room have a safety pin in the center because they do not quite close. To use a safety pin for this purpose is a minor act of invention, and the result becomes a potential ingredient for a larger act of invention. My wife Gloria is sprouting some seeds in a tray next to another window. The sill is not wide enough to hold the tray, so she has moved a step stool to the window and balanced the tray between the stool top and the sill. She then has pulled out the curtain to wrap it around the stool, pinning it closed with a safety pin. In so doing, she has created a miniature greenhouse between the curtain and the window. Notice that this act of invention builds upon the first, using the safety pin to close a curtain. In many ways we invent every day. People are inventive but they do not normally think of themselves in that light. One thing we want to do in this book is to draw out and systematize those inventive tendencies already present in each of us.

Let's begin our study of novice invention with children's inventions and then move to the inventions of some of my university students. The inventions here are concerned with concepts, not the actual hardware. The inventions of both groups are based on keeping an invention diary in which one notes simple human problems that need correcting. It is primarily an exercise in problem finding in which we look for unmet needs. This is just one way to start the process of invention, but it is a useful and accessible one.

15

Suzie Amling was a first-grader and one of the national Grand Winners in a *Weekly Reader* invention contest of several years ago. The contest is a big event, involving hundreds of thousands of children, with competition at the local, state, and national levels. To take up Suzie's story, her class regularly walks two-thirds of a mile along a highway to the library. Suzie is concerned for their safety. Her invention is a Line-leader and Keeper, essentially a long ropelike device with handles along it for children to hold onto. The teacher's front handle lights up and signals the teacher if a child lets go.

Suzie's invention can be thought of in two ways. In the first, the invention starts with a need or functional requirement: to protect the children. Call this type of invention *need driven*. In the second way, invention begins with an existing device or product that does not work exactly right; we then try to modify or improve it. Call this type of invention *device driven*.

When we examine the Line-leader from the need-driven view, we see that its purpose is to keep track of children without the teacher's constantly turning around to look at them. When the children are not where they are supposed to be, when they go astray, their movement is communicated to the teacher. The operations and materials required to achieve this goal are completely unspecified at the beginning of the problem formulation. It is invention out of whole cloth, starting with nothing but the requirement to protect the children. To bring about a tie-in with mechanism, one needs to identify additional requirements and constraints. If the children are to be kept track of without looking at them, then there must an alarm of some sort (either auditory or visual) to indicate when children are not in place. Not wanting to start from scratch, Suzie looks for some existing method of keeping track of children. One of these includes a rope with a series of loops to be held. Such a device, however, does not automatically give notice that children have strayed. How can it be modified to do so?

Notice the progression here. Start with a specified need or functional requirement; identify constraints; ask what kinds of existing inventions will satisfy at least part of that need; and once a candidate invention is found, determine how far short it falls of the newly desired function. Can the rope and loop system be modified to satisfy the needed function? Yes, perhaps, but first a basic decision is necessary. Should the device signal when things are all right or when they are wrong? Everything known about human attention indicates that the signal should be for abnormal or unusual states; the signal should indicate when children stray. What we need is a switching circuit that sounds a buzzer or produces an attention-getting flashing light whenever a hand is removed from the loop. Evidently, this is what Suzie had in mind, although her invention does not specify the details of the implementation; it is at the conceptual level, entirely appropriate for a first-grader, as are the other children's inventions described here. Her invention is a nice combination of new and old things, and it shows a substantial amount of implicit knowledge about how the human attentional system works.

Now let's examine Suzie's invention from the second view, that of device-driven invention. Here we start later in the process, with an existing invention such as the rope and loop system already in use. Then we analyze the rope and loop for what computer programmers call "bugs" or problems. For example, the teacher must constantly turn around to see that everyone is in place. How can it be improved? An answer is to find a way of automatically signaling when a child is not in place. This suggests lights, buzzers, and a switching circuit. Looking for bugs in an existing invention is a common path taken by professional inventors. It also underlies the idea of continual improvement in manufacturing. One doesn't start with a particular functional requirement, only the idea of trying to make something work better in an unspecified way.

Most of the childrens' inventions seem to use the first path, need-driven invention, in which functions are specified and the way to achieve them is unstated. However, the second path, device-driven invention, is often less abstract and seems to involve less memory organization and search. Nonetheless, even the inventions of children are rich enough to strain any simple system of categorization. Let's look at some additional examples, again from the winners of the national *Weekly Reader* contests.

Katie Harding is a kindergarten winner. Tired of splashing herself by stepping in mud puddles at night, Katie invented the Mudpuddle Spotter—an umbrella with a flashlight attached to the handle that helps people see puddles in the dark.

This invention is a combination or *join* of two existing inventions, an umbrella and a flashlight. It offers some advantages over an umbrella held in one hand and a flashlight in the other. First, the umbrella and flashlight come in one package, so it is impossible to forget one component. Second, in the joined form, only one hand is required for operation. The other hand is free to carry things, a considerable advantage. In Katie's drawing, the flashlight is part of the umbrella handle, but I cannot tell from the drawings just how the umbrella and handle fit together. The umbrella still needs to fold, presumably around the flashlight. The flashlight itself needs to have an adjustable angle, so when the umbrella angle is adjusted to take care of wind and rain direction, the light still focuses the right distance ahead. Perhaps the light could be in the butt of the umbrella handle and encased in a flexible cable? Conceptually, I like this invention; we need more kindergartners thinking this way.

Anna Thompson is a sixth-grade winner. Anna needs a way to

quickly and easily measure shortening for her cooking. Her invention is the Measure Quick Shortening Dispenser. Plastic bags that hold two cups of premeasured shortening are fit into a plastic cylinder, and pushing on the base of the cylinder squeezes the shortening out.

Where did this idea come from? A syringe, a cake frosting tool, a caulking gun? Does a cake froster have graduated marking on the side? If so, then perhaps Anna's invention is a generalization of the syringe. Simply consider the syringe's contents and size as variables that can take on any number of different values. What other materials can we use in a dispenser of this type?

James R. Wollin is an eighth-grade winner. James has trouble getting peanut butter out of a jar, particularly the last of it. His invention is the Jar of Plenty, a jar with a double lid—one on the top and one on the bottom. People can now reach the peanut butter at the bottom of the jar simply by opening the bottom lid. Not a morsel is wasted.

It is amazing what the simple repetition of a feature like a lid can accomplish. Who is to say that a jar can have only one lid? For at least some purposes two lids may be better than one. Is it ever possible to have three lids? No lids? No lids will yield a sealed vessel, a particularly useful entity if it is a vacuum tube.

Suzanna Goodin, a first-grader from Hydro, Oklahoma, is a National Winner for grades K to 4. Suzanna needs to feed pet kittens their canned food—and she does not want to wash the spoon. ("Mama yells if I don't wash the spoon. I hate work.") Her invention is the Edible Pet Food Server. Suzanna baked a "spoon-shaped pet cracker, which may be broken up into the pet's food after serving." She "makes her servers in three flavors: charcoal for sweet breath, brewers' yeast for flea control, and garlic to control intestinal parasites."

Suzanna accomplishes several things here by adding some unusual functions to the spoon. Instead of simply being a tool to serve with or to eat with, the spoon is itself edible; an input also becomes an output. Not only is the spoon a food, it also has acquired a putative pharmacological function: it will sweeten the breath, and it will control fleas and intestinal parasites. The efficacy of such treatments is not at issue—they are folk remedies that may not have basis in fact. The imagination behind them is impressive for its combining of complementary functions. We have a spoon made of an edible material (baked pet biscuit). We "dispose" of it usefully, as a food product *and* medication, rather than washing it or throwing it away, as we would with a plastic spoon. Its disposability, in turn, has effected an economy of labor; we no longer need to wash the spoon. Quite a nice combination of properties. Where did such an invention come from?

Actually, Suzanna's concept goes back to a time-honored tradition. An old folk tale of the Southwest tells about a meeting between the Spanish explorer Cortez and the Aztec king Montezuma. Cortez: "I am so wealthy, that I eat only with a gold spoon." Montezuma: "I am even wealthier. I never eat with the same spoon twice." Montezuma then spoons up food with the tortilla, which he eats along with the food. When the food is gone, so is the tortilla. Disposable spoons have a history, and Suzanna's invention extends that history to include the food and medication of her pets.

We often think of inventions as enduring artifacts, but inventions of procedure also are important, and those procedures often have important psychological components. When a problem confronts us, it is often useful to entertain a choice of solutions, ranging from the technological, with an emphasis on artifacts, to the psychological, with an emphasis on procedures. This is not to suggest exclusive alternatives but a sliding scale of tradeoffs between the two.

The Invent America program of the U.S. Patent Model Foundation provides a good example of a problem approached either on the basis of technology and artifact or on the basis of process and psychology. First, the technological approach. A participant in the program, Juan Landas, a seventh-grader, proposed a Velcro glove for catching a cloth-covered football. With the Velcro glove and the cloth-covered football, "butter fingers" become "sticky fingers" and the ball is easier to catch. The Velcro solution to catching the football is largely technological. Something new and concrete is invented or applied in a new way to deal with the problem.

What of a non-hardware approach to the same problem, one that involves inventing procedures instead of artifacts? Can we help a child learn to catch a football using an approach with *no new technology?* Learning to catch any ball is not easy. Many important principles must be mastered: Keep the eyes on the ball; hold the hands together, partly open; face the oncoming ball; don't blink as the ball approaches; and try to anticipate the ball's trajectory by adjusting simultaneously the positions of the feet, body, and hands. When contact is made, close the hands with just the right amount of force and quickness.

One reason that these principles are difficult to master and employ is the time pressure imposed by gravity and the forward motion of the ball; changes happen rapidly. Another reason is that the ball's motion has several "degrees of freedom": it is simultaneously moving in the horizontal, vertical, and depth planes. A procedural approach suggests slowing things down and reducing the degrees of freedom of motion. That way, fewer things need tracking, and there is more time to do what must be done.

Instead of a football flying through the air, we may substitute a round ball that is rolled on the ground or a table surface. The ball's freedom of motion is reduced, and so is its speed. Other intangible factors follow: A reduced fear of being hit in the face contributes to not blinking, to keeping the eyes on the ball, and to spending more time to practice. After mastering ground balls, it's time to move to round balls moving through the air. The round ball is still in order, because it is easier to catch than the irregularly shaped football. By emphasizing procedures with reduced degrees of freedom of motion and relaxed time constraints, we gradually move closer to the conditions of catching a football under real game conditions.

Are the technological approaches centering on artifacts and psychological approaches emphasizing procedures mutually exclusive? Certainly not. For example, a handicapped child, or a child traumatized by failure, might profit from both approaches. The Velcro glove and cloth cover is used along with a round ball rolled on the ground—sticky fingers are combined with a slow, regularly moving ball.

Is the procedural or the artifactual approach better? A general answer to this question is not possible. It depends on the problem, the resources at hand, and the capabilities of the people involved. Usually, the best inventions incorporate both the psychological and the technological. High technology is not helpful if it cannot be controlled by a user, as illustrated by all the complex VCRs that show a blinking ''12:00.''

What else can we learn from this example? Catching a ball has a fairly sharp demarcation between success and failure. Also, a long period of failure may precede significant success. In Figure 2.1 the Initial Period requires up-front effort without much payoff or visible sign of success. In fact, one function of a coach is to provide artificial feedback during this stage by telling the learner when he or she is "getting close," even though the overall performance is still at the failure level. The natural payoff is failure or success, with nothing between. But the underlying performance is continuous; and the continuous nature of that performance needs to be guided during the initial failure period. After some successes begin to accumulate during the Visible Learning Period, performance is likely to become self-rewarding and self-sustaining. Finally, the learner reaches a level of Realized Skill in which performance may still gradually improve, but the improvement is not due so much to learning new routines as it is to adjusting delicate parameters of time and movement to achieve fluent performance.

Tasks with a long Initial Period of failure, such as catching a ball or learning to read, are ripe for inventive minds and inventive processes. Those minds and processes may work at the technological level, at the procedural level, or with some combination or tradeoff between the two. Any device or procedure that produces more early successes has the best chance to succeed as an invention, because it encourages more people to participate in the process. This point is illustrated repeatedly in the history of technology. A good example is the interface between human and camera. It used to be very difficult to take a simple picture. Now, with automatic cameras, it is easy. Another example

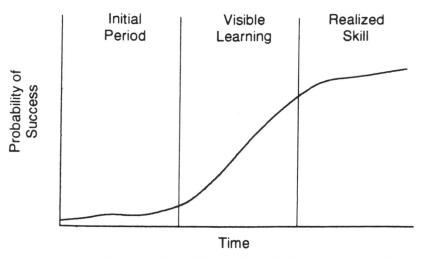

Figure 2.1   Learning to catch a ball: Hypothetical relations between skill and practice level for tasks with an initial period of failure. The longer the Initial Period, the fewer the people likely to persist until a level of Visible Learning or Realized Skill is reached. Such a task is ripe for inventions—like the Velcro football.

is the changing human interface to computers. Computer interfaces using a mouse and menus are easier for people to master than the older command-line interfaces where we are confronted with a blinking cursor on an otherwise blank screen. These two examples suggest a useful rule of thumb for inventors: When doing something is a big hassle, fewer people will participate than when it becomes relatively effortless. To have a product used by more people, improve on the human interface.

The Bugs of Everyday Life

To study invention based on everyday problems, I asked a junior-senior level university class in cognitive psychology to keep a diary of everyday bugs, as a problem-finding exercise. Students had a week in which to note problems as they arose. For each problem, the student was to propose a solution in the form of an invention. The invention did not have to be worked out in detail. These were conceptual inventions—and all that implies. They might be impractical, they might already exist—unknown to the student—but these considerations did not matter. The issue was whether keeping a diary of everyday bugs could suggest interesting potential inventions to students who did not think of themselves as being inventors, many students who if asked would say they could not come up with any original ideas at all, let alone an invention.

The instructions were as follows:

> Your job is to start collecting information on some of those bugs in your own life. For each bug, try to sketch out what it would mean to have a satisfactory solution and, if possible, the type of invention that might implement that solution. Consider an example. I am always losing my car keys. Solution: My car keys will beep or blink on and off if they have been lost. To accomplish this requires some sort of Invention: Perhaps I can use a scanning beam that will shine around the room to activate the beeper or blinker on my key ring. The remote infrared controller from a TV set might function as the beam; it is available, if I can find it. Or several hand claps in rapid succession might start the beeper on my car key ring.

Here are some results from my students' invention diaries. Most of the inventions are not really original in a strict sense. However, they were original for the student, and certainly there is something to be said for the independent rediscovery of an idea. This was a pedagogical exercise; it served to stimulate thinking about invention, and it helped students see the scope of their invented world.

John's problem is not knowing how much his electricity bill will be. His solution is to have a running total displayed right underneath his thermostat. The invention then is a meter that will give the running total in dollars for the month, so he can budget accordingly or try to limit his usage. I think this is a great idea for consumers, but the power company may not like it! Certainly, it is well within the realm of current technology to have such a device. It should be adjustable for the rate in a given area which is multiplied by usage to give the monthly bill to date, as well as the projected bill based on current usage. Why use dollars instead of kilowatt hours? Probably because there is more motivational power in money than in kilowatt hours. The meter should be inside so that it can be conveniently monitored near the site of the crime (the thermostat), just like bathroom scales really belong in the kitchen next to the refrigerator.

Lisa's problem is that during winter her car is all iced up and cold. Her solution is to have it start ahead of time, so it would be warmed up before going out. Her invention is a remote controller like TVs have that would start it. Of course, some safety problems need to be solved, such as starting only if the car is in neutral and the brake is on. But problems like this should not be insurmountable. Cars ought to be at least as intelligent as TV sets. This was an original idea for Lisa, but beyond the context of the class it is not original. Expensive versions of such devices are now on the market for another reason; people who think they may be the target of a terrorist bomb can start their cars remotely.

Tom's problem is waking up in the morning. He hits the snooze button several times, but keeps going back to sleep. The form of his solution is some way of gradually waking him up. His invention is a time-release pill that will slowly wake up his system. For example, if he wanted to sleep for eight hours, he could take an "8-hour pill." After sleeping 7½ hours the pill would begin releasing caffeine, and by the time he had to get up, he would feel like he had already had his first cup of coffee and wouldn't have to postpone (because of lack of time) his shower. This sounds as if is technically possible now. One of the big breakthroughs in pharmaceuticals has been the *time-release* capsule, one that emits a steady release over time. This is a bit different; it is a *time-delayed* capsule that does not start its release until 7½ hours later. Probably, the pharmaceutical companies would be delighted to have such a product, because of its money-making potential. I do have some reservations about turning over large parts of life to chemistry, but the concept is an interesting one.

Jane's problem is locking herself out of the house. Her solution is to have an electronic scanner that can read her fingerprints. If the fingerprints matches, the door will unlock. Her invention then is a fingerprint reader and lock. Such inventions are now being worked on by the military for reasons of computer security and for entry to critical installations. A biologically based security system offers advantages. The "key" is always with you and cannot be lost; it is also difficult for other people to duplicate. But probably not impossible.

So what have we learned from our novice inventors? Several important ideas. Invention is not a rarified activity undertaken only by the genius in the laboratory. Instead, when given the chance, it can be an everyday activity for each of us. Next, a good place to begin an understanding of underlying principles is with children and older amateurs. For these groups the ideas are not yet complex and the processes behind them are still reasonably simple.

But there are limits to what simplicity can teach us, so it is time to go to the work of the expert inventor. However, to retain some simplicity, we will follow a common, everyday idea as it winds its way through history.

3

Expert Invention and the Turn of the Screw

I'm fastening a shelf on the wall. I want it to be a strong enough for books so I'm using long wood screws that will not pull out. I've never looked closely at a screw before, but now that I've become interested in invention every device seems new to me. Where did the screw come from? From some natural form? Or the less complex nail?

While we all engage in invention on an everyday basis, with the motivation of satisfying an immediate need or making an immediate improvement in the way something works, that is not necessarily the way the experts do it. To be sure, the expert's motivation may well be need-driven or device-driven, but it may also be something more. For lack of a better phrase, we will call that something more *play-driven* invention. It results from the actions of mental manipulations driven more by mental play and the desire to form interesting combinations than anything else. It is *what-if* thinking on a grand scale—thought experiments where ideas and devices are selected and combined in unusual ways. The combining of ideas is anything but random, however. Often it is influenced by subtle and powerful heuristic principles, rules of thumb for generating ideas. To foreshadow our story, I want to follow the idea of the screw through the work of great historical inventors—Archimedes, Leonardo, and the Wright brothers. The span of history ranges from the third century B.C. to this century, and each case is a meditation, a variation, on the turning screw.

A Twist of Archimedes' Mind

In the American Midwest a common activity is unloading grain from trucks into one of those most original of architectural forms, a grain elevator. The unloading is accomplished with a grain auger, a drill bit covered with a cylindrical sheath. As the bit turns, it pulls the grain up into the recesses of the elevator. These augers were first used not for grain but for moving water.

In the third century B.C., Archimedes (c. 287–212) invented the water screw, a turning device that gets water to flow uphill. He is surely one of the great intellects of history, someone who made many original mathematical contributions and provided the laws of the lever. Figure 3.1 shows an illustration of the water screw. Unlike the contemporary woodscrew, the water screw is fixed in position longitudinally as it rotates, like the tuning peg of a guitar. It is covered with a wooden casing or tube, and the encased screw enters the water at an angle. When a handle at the top end is turned, water "runs uphill" and empties out at the top into a trough. The water screw's capabilities made it useful for irrigation or for draining seepage from mines. It is still used in the Middle East to lift water. And it is the basis for the old hand-operated food grinder that moves solids through a rotating coil.

When thinking about the water screw, some people have trouble envisioning water running uphill, but it does. To make it easier to understand, think of your old food grinder with an augerlike screw, and

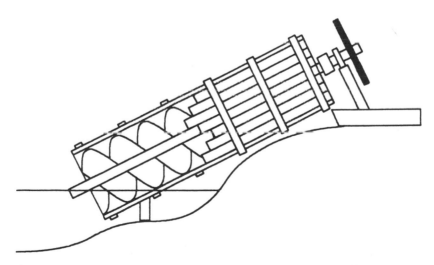

Figure 3.1   A cross section through the water screw. The exact form used by Archimedes is evidently not known. For the actual water screw, the fit between the auger and the casing walls must be tight to keep the water from leaking down. (Drawing by M. Sheldon, after Drachmann, 1967.)

similarly tilt it. As the crank is turned, the food moves between the threads of the screw and it *will* go uphill relative to the length of the whole screw; the water movement in Archimedes' screw takes place in the same way.

The problem to address has a surface simplicity: How did Archimedes get the idea for the water screw? Of course, we will not be able to give a firm answer based on the historical record; it is much too skimpy. Instead, we will try to provide a plausible account. The validation of such an account is then determined by seeing if we can extract underlying principles for invention that have applicability beyond the particular case of the water screw.

To really understand the water screw as well as the other screw-forms that will be introduced, we need to understand one of invention's grand ideas: We can make tradeoffs among effort, space, and time. The inclined plane is an apt example. To get to the top of a tall building without stairs or a ramp, we could simply put on a shirt with a big "S" and leap to the top in a single bound, if our muscles and gravity allowed it. Unfortunately, muscle and bone are not up to the task because of the power requirement—too much work per unit time for leaping. But with stairs and ramps we are able to get to the top of the same building by covering a longer distance and taking a longer time. The advantage in doing so is that at any point a more manageable effort is required to climb the stairs than to leap to the top. We trade off space traveled and time to get to our destination against the peak effort required. All of this is made possible by the inclined plane (illustrated in Figure 3.2).

With these ideas in mind, we are ready to return to our question: Where does the idea for the water screw come from? A need-based explanation has Archimedes wanting to get water uphill to animals or crops. He then searches his memory for a suitable mechanism, and perhaps he recalls his outings to the beach as a child. Suppose he played

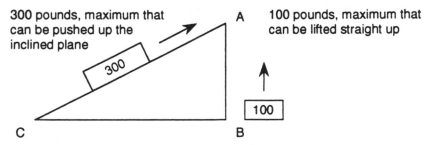

Figure 3.2 The inclined plane and mechanical advantage. Distance and effort are traded-off against one another in lifting a weight. Instead of the brief intense effort of pulling straight up, the inclined plane offers the option of a longer but less intense effort distributed over time.

with cone-shaped aquatic mollusks on the beach, the kind with coiled, screwlike shells. Each of these is actually a long narrowing tube wrapped as a helix. If one of these was broken at the pointed end, and was turned in the water, Archimedes could have discovered that the water came out of the top. Without realizing its significance, he had the equivalent of the water screw. Much later, as he began to think about the problem of getting water uphill, that childhood memory returned to him. This account is a biological model of the origin of an invention, a found-in-nature explanation, offered after the fact.

Unfortunately, it is an account unrelated to other aspects of Archimedes' thought; it does not build on any of his prior work. Moreover, as the historian of technology Bertrand Gille has pointed out, the water screws of Archimedes' time were based on a cylinder rather than the cone shape of a shell; and they did not use a coiled tube in their construction. It was not until 1600 years later that Leonardo da Vinci employed a coiled-tube variation on the water screw. For these reasons, a biological model is an unlikely source for the invention of Archimedes' water screw.

A *found-in-nature explanation* like this one is a recurrent basis for explaining invention origins. Comparable accounts have been given for the origins of inventions as diverse as mythological forms and Velcro. But such explanations often seem as though they've arrived from left field. And even though a found-in-nature explanation may be historically accurate for some inventions, it is not readily generalizable—it won't apply to other situations. It tells us little about invention, other than to look closely at the world about us as a source of ideas, a vague bit of advice. Even when there is evidence that a natural form was the inspiration for an invention, we still need to seek other possible origins and relationships, because many paths may lead to the same invention—and our purpose is to provide a systematic account. To be systematic is to seek principles and generalizations behind the appearance of invented forms and to make the discovery or invention process more rational. If we can find powerful principles behind one path, those principles may have broader applicability for paths to other inventions.

A second explanation for the origin of the screw is device-based; it comes from simple mechanical devices used in practical construction settings and then generalizes on them. In fact, the screw may be regarded as a generalization of the inclined plane or ramp. We simply wrap an inclined plane around a cylinder and end up with a screw. But this leaves many important steps to fill in.

Certainly, inclined planes with a twist were used in construction long before Archimedes. Some authors believe that the pyramids were constructed this way: Lay a level course of stone blocks and then build an inclined plane or ramp of earth up to the top, lay the next level of blocks, and so on. As before, the purpose of using the inclined plane for raising construction materials is to trade off the total distance covered for a more gradual ascent and less peak effort. The inclined plane wrapped around a four-sided pyramid does not produce a circular form such as a screw, but it is a simple next step to wind the plane around a cylinder. Seen in this light, the screw is the result of refining a construction ramp. However, this is a long way from Archimedes' screw. The construction ramp is fixed in place and solid materials move over it.

Intuitive construction techniques with inclined planes may be one precursor to the water screw, but Archimedes was certainly capable of reasoning through the process using a kind of mental play not specifically driven by practical needs. Let's see how play-driven invention might work. The key feature of the inclined plane is the mechanical advantage it presents. In Figure 3.2 we consider 100 pounds as the maximum that we can lift straight up, from B to A. However, if we pull something up the incline C to A, the maximum weight is 300 pounds: the length $CA = 3 \times AB$ for a mechanical advantage of 3 to 1.

For a fixed height, the greater the length of the incline, the less effort required at any given point in time. This idea, together with that of a winding ramp placed around a pyramid or other building, suggests the desirability of *packing* a long incline into a small area. The principle we are searching for is how to take maximum advantage of the inclined plane in a minimum area. The appropriate packing principle is to coil the inclined plane that covers a fixed area into a third dimension, so minimal area is required. The result is a screw.

The packing principle provides the greatest possible length of inclined plane—and the greatest mechanical advantage—over a fixed area. All of this is achieved by escaping into another dimension and winding the inclined plane above itself. The result is now a screwlike form. It is not the result of generalizing on a construction technique, but is instead the consequence of an abstract exercise for packing a geometric form into a small area. While the result looks like a screw, the story is not over.

How do we get beyond mere shape to the different idea of a *water screw* that will lift a liquid? We see that the water screw differs from a ramp in important ways, other than its construction. For the inclined plane used in construction, solid materials are moved up a fixed or motionless ramp by people pulling or pushing. Suppose we vary the critical parts of this assertion: The plane now moves past the materials and the material is changed from solid matter to liquid. We have then generalized on the materials moved, going from a solid to a liquid. And we have reversed what moves; instead of having the materials move over a fixed plane, the water screw itself moves rotationally as a coiled inclined plane through the medium of water. We can think of these inventive moves as involving a reversal principle: If a device like an inclined plane has material moving over it, try the reverse process; move the device past or through the fixed material.

When we combine these principles, they suggest the possibility of moving water uphill with a moving screw. However, other problems remain. The water will yield to gravitational forces and fall from the rotating screw; some force must be found to turn the screw with good mechanical advantage. The solution for the falling-water problem resides in discovering other ways of containing water. The connection or link is found in tubes and pipes, which also move water and contain it. Thus Archimedes' solution is to join the idea of a tube with that of a screw, that is, to cover the screw with a casing. That way the water will not fall out as the screw turns. In fact, the water screw of Figure 3.1 is covered with a wooden casing. The solution for the second problem, an adequate force for turning the screw, is found with some form of handle at the top, a special case of the lever. (I have not used a crank here because that was a later development.)

The combination of parts just described almost solves the problem of getting water uphill, but not quite. If Archimedes' screw is inserted vertically into the water, ordinary hand power may not be able to turn it fast enough to "pull" water up, because of the low friction coefficient of water. But the inclined plane can help us again. Simply tilt the screw into the water, so that water is trapped in the rotating chambers of the screw and casing and is pushed inexorably upward with the turn of the handle. We see that the principle of the inclined plane is used in several ways: first to create the screw form and then to tilt it into the water.

Of course, any of the accounts—need-driven, device-driven, or play-driven—could have produced the water screw. It is a truism that multiple paths can be found to the same invention; that makes reverse engineering possible. I would lay my money on the last account, however. Everything we know about Archimedes indicates a person capable of playing with ideas in unusual and wonderful combination. If the play driven account is right, we have then retraced the masterful mental journey of Archimedes from inclined plane to water screw.

I have presented this story to students and colleagues, and a frequent reaction has been that it is interesting, but how did I come up with it? The steps are as follows.

First, I parsed or decomposed the finished water screw into components: screw, casing, handle, material moved. Second, I parsed these components still further. The screw was decomposed into an inclined plane; the inclined plane was noted as having motion; the material moved was water; the impetus for turning the water screw was the handle, a special form of lever with the fulcrum at the center. Third, I asked what principles would reverse the direction of description, moving from whole to part. For example, I wanted to know how to get from the inclined plane to the screw. The appropriate reversing principle involved packing the plane into as tight an area as possible to increase the mechanical advantage through a long run of the plane in a minimum area. Fourth, a big and important step was to treat the materials moved as a variable, something that can change. This suggests not only solids like rock and sand to be moved, but also liquids such as water. Fifth, I tried to figure out why the casing must be joined with the screw. The answer involved gravitational pull on the water, together with water's slippery propensity for running downhill. The inadequate screw action and an enclosing casing were linked by their common purpose, carrying water. They formed a complementary unit. The inclined plane principle of the screw gets things uphill and the casing that carries water keeps it from falling out. Sixth, I asked what other ways will keep the water from running downhill. The answer is another use of the inclined surface; this time the screw enters the water at an angle. Finally, I questioned the function of the remaining major part, the handle. Obviously, it was concerned with providing torque or twisting force required to turn the screw.

Notice that all these steps involved a *backward search*—finding a reason for the use of each component and its relation to the other components. Archimedes was confronted with the much more difficult task of *forward search*—finding parts and relations between parts that would do what was necessary, run water uphill. The nice part of backward search is that it allows us to quickly gain insight into existing inventions and the underlying principles that seem to drive them.

A Little of Leonardo's Mind

Let's continue to pursue the turn of the screw, this time through the mind of Leonardo da Vinci (1452–1519). What would happen if the inclined plane, on which the screw is based, did not transport either solids or liquids? What if it were concerned with a gas such as air? To examine this audacious generalization, let's look at Leonardo's helicopter. A slightly simplified version of his helicopter sketch is shown in Figure 3.3. Examination of the figure reveals that Leonardo's helicopter is *an invention before its time*. To be successful, he would have needed light materials, a small powerful engine capable of lifting an air vehicle and its occupants, and a requisite knowledge of aeronautical principles. These critical ingredients would not be available for another four centuries. Nonetheless, the essence of "helicopterness" is apparent. Again, our task is to construct one or more plausible accounts of how such a dramatically original idea came to pass. The task is daunting; evidently Leonardo had little to say about it beyond the drawing and a few accompanying words:

Let the outer extremity of the screw be of steel wire as thick as a cord, and from the circumference to the centre let it be eight braccia. [4 meters] *[Unlike Archimedes' longitudinally fixed screw, this one can move through a medium, air.]*

I find that if this instrument made with a screw be well made—that is to say, made of linen of which the pores are stopped up with starch— and be turned swiftly, the said screw will make its spiral in the air and it will rise high. *[Once again, the screw moves through the medium of air; weight reduction efforts are also made through the use of a fabric.]*

Take the example of a wide and thin ruler whirled very rapidly in the air, you will see that your arm will be guided by the line of the edge of the said flat surface. *[A small physical model is used as analogy.]*

The framework of the above-mentioned linen should be of long stout cane. You may make a small model of pasteboard, of which the axis is formed of fine steel wire, bent by force, and as it is released it will turn the screw. *[Further weight reduction is emphasized through a skeletal structure with a fabric covering; a small model is constructed; a spring may be intended here as one source of power.]*

Admittedly, the quotation does not give us much detail, and the drawing also provides its own ambiguities. In the original drawing there are lines that look like support wires or projection lines. Without affecting the argument here, we shall assume that they are projection lines. However, several things are clear from Leonardo's drawing and text. The helicopter is propelled by an air screw. Weight is recognized as a problem, and attempts are made to minimize it by using a light construction of cloth and cane. The idea of lift is introduced by the model of a ruler whirled through the air. Even though the figure and the quotation leave out important information, we can construct several paths along which Leonardo's imagination may have traveled. By doing so, we will illustrate principles and methods of invention that will make the process less mysterious, even though we may not become new Leonardos.

To begin, can a need-driven explanation account for the idea of the helicopter? I don't think so. No one needed to fly in any obvious sense. However, people have certainly wanted to fly for a very long time. Starting with that want, a kind of connection can be made with existing objects capable of flight. The bird suggests itself; people see it and they also want to fly. In fact, Leonardo did base other flying machine sketches on a biological analogy, and all that implies with wings stretched wide and flapping. The bird as a model of human flight goes far back in history to dimmer visions than those of Leonardo. It occurs in Greek mythology with joined human and animal forms, and with Daedalus and his son Icarus. Remember how they escaped the labyrinth of the wild bull-like Minotaur by using hot wax to fix bird feathers to their bodies. Then they flew out of the maze. So enamored of

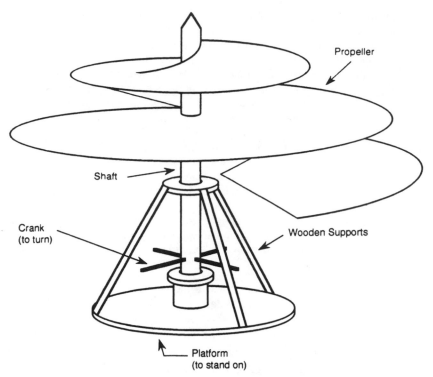

Propeller

Shaft

Crank
(to turn)

Wooden Supports

Platform
(to stand on)

Figure 3.3   Leonardo's helicopter. This is a conceptual invention, one well before its time because it could not possibly have worked. Nonetheless, it is a grand idea. (Drawing by M. Sheldon and M. Ast.)

flight was Icarus that he committed hubris; he acted like a god by flying too high and too close to the sun. The wax that had yoked feathers to his body then melted, and he fell to his fate.

The bird and its feathers are natural associations with flight; but the helicopter is not. It is based on the screw, a different and more radical principle, one that strikes out not along the path of flesh and feather but along the path of mechanism. People do not see a screw and then want to fly. If Leonardo's helicopter is an idea not based on need or wanting to be birdlike, where did it come from? I can think of several different explanations.

The first is a device-driven one, in this case based on an analogy to an ancient Chinese twirling toy that Leonardo may have known. I have not been able to find any pictures of this toy, but whatever its nature, it is scaled up in size and the idea of a vehicle and passengers are added. This may be the easiest route to Leonardo's helicopter, but we know little of its background, nor is there firm evidence to indicate that Leonardo's idea derived from it. Like a vague found-in-nature explanation, this one does not build on his prior work so we will not consider it further.

Let's try another mental path, one that does consider Leonardo's prior work and shows how to work from a model. What if he started with a screw press for olives, as Figure 3.4 illustrates. The Greek Heron described such a press in the first century A.D., and it probably dates to the first century B.C. A screw made of wood, with a lever handle on top, penetrates a threaded wooden frame and drives a platform down onto olives in order to squeeze and extract oil. Leonardo was certainly familiar with such an olive press; indeed, he used something like it in a design for his printing press, which in form resembled the Guttenberg press that may have earlier been derived from a grape press.

Given an olive press, how can we generalize from it and arrive at the helicopter? Well, the platform functions a bit like a vehicle, because it moves through space, but, in this case, up and down. Suppose it were a vehicle, what modifications would make it work better? Given

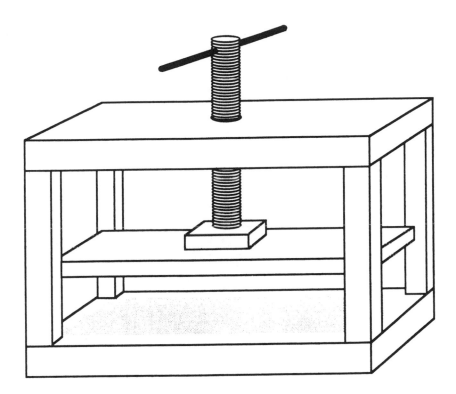

Figure 3.4 An olive press as a possible model for the helicopter. When the crank is turned, the screw of the olive press moves through the medium of wood. When the crank of the helicopter is turned, the screw moves through the medium of air. (Drawing by M. Sheldon and M. Ast.)

the up-and-down movement, a major insight is that the screw, instead of moving through the press's wooden frame, must move through air. To do so, it requires reshaping and enlargement for added lift. Then we come to the source of propulsion. Clearly, the turning handle for the screw press must be translated to a lower point on the device, so that it is accessible to manipulation by the passengers on the platform, and the number of handles should be increased for added force and lift. Unnecessary parts of the olive press also need to be deleted, things like the base and the frame. The result is that by changing the scale and rearranging different parts of the olive press we have transformed it into a helicopter much like Leonardo's.

Lots of mental play is involved in moving components around and seeing what is possible. And we can very well end with the same result by a different pattern of play; usually a given invention has more than one path to it. To see this, let's now work not from the olive press and the screw form but instead start from the idea of a vehicle. For Leonardo, existing vehicles included boats and carts. One moved through the medium of water, the other over land. One was propelled by wind or oar, the other by horse or ox or human. The fact that boats and carts already moved through different media or spaces suggests that the type of space through which a vehicle moves can vary. Indeed, another possible variation is air. A biological way of moving through air is to follow the model of birds, with feathers and flapping wings, as we already considered. Yet another model for moving through a medium is the oar of a boat. But that just does not seem right for the air.

What other mechanical devices move through media? A knife, based on the idea of the wedge; but what will thrust it through the air? The nail, also based on the idea of the wedge; but what hammer will pound it through the air? The screw, based on the idea of a coiled wedge or inclined plane; but what will turn it through the air? Unlike knife and nail, which penetrate only through linear motion, a screw works through rotary *and* linear motion. Continuous rotary motion is possible, provided by a crank-driven shaft or a coiled spring, and with continuous rotary motion the screw produces continuous linear motion through a spatial medium. Perhaps, the screw is just what is needed for propulsion. Notice that the link among vehicle, flight, and screw can be made by starting with the vehicle and moving to the need for continuous rotary motion, and then finally to the screw.

To follow this mental path in more detail, plainly the screw must turn in the medium of *air* and move through it. This departs from Archimedes' screw, which only moves rotationally in the medium of *water*. Instead, the screw form most relevant to the helicopter is the wooden one used in grape and olive presses to extract juice or oil. There is another distinction between Archimedes' screw and the screw of the olive press. The first rotates in a fixed place and moves the medium (water) past it; the second both rotates and moves through a fixed spatial medium (wooden frame). The screw of Leonardo's helicopter is more like the olive press screw; it moves not through a wooden frame but through a gas, the medium of air.

The combinatorial play of screws in Leonardo's drawings is impressive indeed. There are fixed screws and moving media, as with worm gears; and fixed media and moving screws, as with the olive press. Also included are generalizations on media moved through: wood (presses), metal (screw-making machines), and air (helicopter). In addition, there are screws combined with gears (worm gears); and screws driven by hot air from a furnace to automatically turn a spit. The last item, the air-propelled screw, is essentially a complement of the helicopter's screw.

How did the idea of vehicle marry with the olive press? A characteristic of Leonardo's drawings is the fluidity of the size dimension. The drawing mode is intrinsically flexible in scale as compared to a construction with real objects. In a drawing, a screw or other device can readily appear as large, or even larger, than a person in another part of the same page, a common occurrence in Leonardo's notebooks. So the physical difference in size between the screw of an olive press and that required for a helicopter's screw rotor may readily disappear in the drawing and also in the corresponding mental representation of the two. It is then not a huge leap to get a man on board that moving screw, especially when the olive press already has a platform that can serve as a place for passengers.

In our speculative journey through Leonardo's mind, we have put different ideas together. The helicopter is the combination of the olive press screw, a land vehicle, and the idea of flight—all with a number of scale changes. We do not know that Leonardo ever thought this way. But he was fascinated with screws of all kinds, with presses, and with the problem of flight. What we have done in our account is to put all three together.

Notice that invention as systematic mental play involving abstraction and generalization may be strikingly different from the idea of invention based on necessity and concrete problems. Both approaches undoubtedly take place. But lay people—and many engineers—ground invention solely in need-based problems. That is an undue limitation. To also include abstraction and generalization-based invention rooted in mental play is to move toward the approach of mathematicians who freely surrender to the allure of magnificent mental abstractions and the pursuit of a form for its own reasons.

But the account here may appear too facile. Certainly, there is more to invention than simply combining ideas such as a vehicle and the screw of an olive press. Much remains. How to fit the parts together;

how to support things; how to keep the load or people from twisting with the screw; and how to propel the screw. Indeed, it is on just such points that the impractical nature of Leonardo's brilliant helicopter concept is revealed. A helicopter propelled by people pushing a crank around a platform would not lift off because of Newton's law (for every action there is an equal and opposite reaction). If the power were adequate, the helicopter would twist in the air; that is precisely why modern helicopters have an additional side-mounted propeller at the end of the tail—to prevent twisting in flight. Also, the power provided by several people turning the crank is certainly not going to compensate for the weight of machine and people, so not enough lift will be available to get the helicopter off the ground. Having a satisfactory solution to such problems distinguishes a workable invention from just a concept. Leonardo was a long way from a truly workable helicopter.

Nonetheless, he did have the helicopter idea. Did it really come from an olive press? I have offered this path not because of its certainty but because of its possibility. Other variations on the screw, such as Leonardo's air-turned cooking spit, could have been put forth. The key idea is this: It is not necessary to offer ad hoc explanations, like seeing a toy or finding something in nature, to explain the origins of the helicopter. The idea naturally arises from Leonardo's own work, if we apply a few powerful principles of invention such as generalizing on the screw and the media through which it moves. As with found-in-nature explanations, ad hoc explanations based on toys may turn out to be not only dubious but unsatisfying.

By looking further, we see still more lessons revealed by Leonardo's helicopter. Based on the advantage of hindsight, we can discern several significant steps that Leonardo did not take. If he had oriented the screw axis of his helicopter horizontally, he would have had something approaching a modern plane's propeller, providing for lateral movement through the air. From his study of bird flight, we know that Leonardo was in fact interested in lateral movement through the air. He has elaborate sketches of mechanical wings to be moved by human power. A screw moving horizontally through the air is more like bird flight than the up and down helicopter, and simpler as well.

Still another step was not taken. If Leonardo had oriented the screw shaft horizontally in water, it could have powered a boat. In fact, sailors refer to a ship's propeller as its *screw*. All he needed was for the helicopter screw to be given a ninety-degree rotation for lateral movement through the medium of water. In fact, Leonardo often drew analogies between air and water, but he did not seem to apply the idea of the screw to water. While a successful helicopter or aircraft was not possible in Leonardo's time, the use of the screw in water as a propeller may well have been workable, but it did not come to fruition for another three centuries.

Even though opportunities for generalization were missed, Leonardo's helicopter was an idea and a machine before its time, a compass bearing into the future.

The Wright Stuff

On December 17, 1903, on a remote beach in North Carolina, Wilbur and Orville Wright produced the first powered, controlled, sustained, and manned flight of a heavier-than-air craft. For thousands of years humanity had dreamed of human flight. The dream was first realized not in Washington, New York, London, Berlin, or Rome; nor by sophisticated engineers or scientists. Instead it was realized by two bicycle mechanics, Wilbur and Orville, from Dayton, Ohio. It was as if the problem of nuclear fusion had been solved by two guys from Bug Tussle, Oklahoma. The Wright brothers' feat is one of the most audacious in the history of invention. To understand it, once again, we follow the turn of the screw.

The Wright brothers divided or parsed the problem of flight into three parts: lift, control, and power. This is abstract conceptualization, far from the immediacy of need-driven invention; in short, it is mental play of the highest order. I think this initial parsing of the problem was one of their most significant contributions. If a complex problem can be satisfactorily divided into parts, then the complexity is greatly reduced because each part of the problem can be attacked separately as a smaller problem. But typically a complex problem can be divided any number of ways. Why their particular division? A good parsing is along those natural lines of division that make a problem's parts *independent* of one another. When the parts are independent, what we do in subproblem A, as long as it is effective, does not introduce new complexities for subproblem B. Further, when A and B are solved, their respective solutions can be fitted together, with only minor adjustments to the overall configuration. That is the ideal. Unfortunately, problems do not always present themselves with clear divisions or parsings. Often, those divisions must be hewn from granite with much backtracking. Nor does the best parsing inevitably yield independent components. We hope at least to identify semi-independent components—which is what the Wrights achieved with their three subproblems of lift, control, and power. Let us see what they did in each of these areas.

We begin with the problem of lift. When the Wrights decided to try their hand at flying in 1899, much had already been accomplished. Hardly anyone was still working with flapping wings. Students of flight had begun to speculate about optimal wing shapes, the wind tunnel had been invented, and numerous publications on flight had appeared.

One of their early sources of inspiration consisted of tables on wing lift and surface area that the German glider pilot, Otto Lilienthal, had compiled. Using the values from those tables, the Wrights built what they hoped would be a manned glider. But even when their glider was launched down a slope into a high wind, it did not have the lift predicted from the tables and they were forced to fly it as an unmanned kite.

The problem of inadequate lift continued to plague the Wrights. They developed a distrust of Lilienthal's tables and decided to get their own data on lift. One of their first attempts to secure that data used a V-shaped bracket mounted horizontally on a vertical shaft. Different wing shape pairs were then fixed on the arms of the V and faced into the wind to determine which member of the wing pairs had the least wind resistance and the most lift. This was a crude method since the wind came and went and shifted directions. A next step was to horizontally mount a bicycle wheel on top of a bicycle's handle bars. Then, on a windless day, the cyclist would fix alternative wing shapes along the wheel rim. As he moved, he determined which shape provided the greatest lift and the least wind resistance. The riding-along method must have made it difficult to change variables and make measurements.

The final step was to construct a series of wind tunnels of increasing size. A wind tunnel is essentially a long box into which a fan blows a controlled wind. In the wind tunnel, the different wing, tail, and propeller shapes can be tested for their aerodynamic properties. Specifically, the Wrights were able to measure lift, wind resistance, and other variables while they systematically changed wing, tail, rudder, or propeller configurations. A common method was to suspend or mount a given component in the tunnel, attach a string to it, and then feed the string to a scale that would measure the forces exerted.

While the Wrights were not the inventors of the wind tunnel, they were probably the first to exploit its potential for careful measurements of systematically changed variables. The tables for lift and the other variables they compiled from their wind tunnel work then became the new foundation for their aeronautical designs.

The next major part of the problem to consider is the control system. Early on, the Wrights concluded that most gliders, including Lilienthals', were not stable enough for safe flight. Indeed, Lilienthal himself was killed in a glider fall. The instability problem suggested a need for control. In flight, three major axes of movement must be controlled: pitch, yaw, and roll.

Pitch refers to the up-down motion along the front-back axis of the plane. The Wrights controlled it through the use of a horizontal elevator (a small wing form), called a canard, mounted in *front* of the wing. By changing the incline of the canard, the pilot could raise or lower the angle of incline. The mounting of the canard on the front probably was done for ease and rapidity of control, but its placement was fortunate for other reasons as well. It provided greater stability because a dive would tend to lift it and thereby restore level flight. Also, the front-mounted canard offered greater safety; during the several crashes, it probably protected the pilot.

Yaw refers to a horizontal sliding motion, much like a partial crab walk. The Wrights controlled it with a vertical tail rudder. As Tom Crouch, the historian of flight, points out, neither the pitch control of the horizontal canard nor the yaw control with the vertical rudder were original to the Wrights. The rudder idea comes from ships and the canard elevator is probably a generalization on the same idea to the up-down dimension in air.

Roll refers to rotation about the front-back axis, and the method of controlling it is original to the Wrights. Their analysis of other investigator's problems convinced them that roll must be controlled for greater stability and allowing turns to be made. The prevailing thinking was that turning could simply be accomplished by moving a rudder, much like turning a boat. But for various aeronautical reasons this is unsound and may lead to stalls. The Wrights settled on a method of roll control called "wing warping." A network of cables operated by the pilot would pull down the wing tip on one side and lift it on the other. The effect is to make the entire wing into an adjustable *aerial screw;* depending on the direction of the wing twist, the plane will bank and turn one way or the other. The Wrights' idea for wing warping may have come from observations of buzzard's wing tips while soaring and banking, a biological analogy. Or it may have come from rotating a long twisted box, an idea closer to a screw. The banking effect is also much like that encountered in a bicycle turn, when we lean into the side we are turning to. Thus bicycle turning also serves as a partial analogy. In any case, the actual implementation of wing warping is similar to that of a screw with adjustable threads moving through the air. Wing warping was the idea that the Wrights patented.

Their human interface to the different controls varied. The front elevator or canard was controlled by a hand-operated lever. The wing warping was accomplished once again with a movement like that of a bicyclist's. The pilot laid prone, stomach down, across the wing. His hips were in a movable cradle attached to control cables. By sliding into the direction of a turn, the cradle moved and pulled the cables to warp the wings and bank the aircraft in the direction of the turn. This is a clever interface; it frees the hands, it draws on natural gravitational forces, and it resembles a bicycle model. The rudder was controlled differently; first, it was fixed in position, and later it was made adjustable. When adjustable, changes were first made by hand; later on, linkage with the wing warping was provided; and still later, control was by hand again.

An issue that sharply divided inventors of flight was that of automatic stability versus manual control. Prominent aeronautical investigators like Octave Chanute and Samuel Langley wanted automatic stability—that is, they wanted the aircraft to move in a straight line at a level altitude, if the hands were removed from the controls. Automatic stability sounds like a good idea; it would be the "fail-safe" approach. Langley wanted it because he believed that the air above the ground was turbulent and treacherous; human reactions were not fast enough to deal with the turbulence. However, the Wrights explicitly rejected this notion. They knew a great deal about turbulence. However, the Wrights explicitly rejected this notion. They knew a great deal about turbulence from their kite-flying experiments and perhaps they were again aided by knowing that a bicycle is not automatically stable. They insisted on manual control, although they were willing to link wing warping and rudder movement. In retrospect, automatic stability was a premature and unnecessary constraint that imposed too much complexity for the time.

The final problem to examine is propulsion. Only after the Wrights found solutions to the problems of lift and control, and had learned to fly gliders as well, did they consider adding a motor for power. They made calculations based on weight and lift to determine the power needed. They first tried to buy an existing motor, but found that one meeting their specifications was too expensive. Accordingly, they built their own. It was not state of the art, and it was not powerful in comparison to motors used by Langley. Still, it did not matter, because they knew the power they needed and could build a motor to meet their specifications.

A related issue was propeller design. What shape should a propeller have? They intended to rely on information from the design of *ships' screws,* a reasonable analogy on the face of it, but they found that little theory was available. Because their resources were limited, they did not want to try an extensive trial-and-error search. In thinking about the problem, they decided that a propeller was not so much like a screw, which was the prevailing view, as like a rotating wing. Already, they knew a great deal about efficient wing design. What remained was to calculate the speed of the propeller at different distances from the center shaft and then apply existing knowledge about optimal wing shape and lift at different speeds. The result was a highly efficient propeller.

Notice the change in analogies here, from the screw to the wing. This was an inspired mental link because there was no theory of screws moving through a medium, and they already knew how to design wings.

In their design of propellers, the analogy of the screw is finally out-grown and it is replaced by another, that of the wing.

Notice that the Wrights' solution to the problem of flight involved contributions to the three fundamental subproblems, aeronautical de-sign for lift, the control of motion, and finally propulsion. Not only did they make contributions to each area, but they meshed the three to-gether in a seamless way. Since these problems were not completely independent, the smoothness of the seam between them is what in the end really made flight possible.

Tom Crouch asks why the Wrights were the first to fly a powered, controlled aircraft. It is a good question because every prevailing stereotype suggests that it should have been someone else. To answer the question, let's examine some contrasts with the work of a compet-itor, Samuel Pierpont Langley, the general secretary of the Smithson-ian Institute and an ardent pursuer of human flight.

First, consider background factors. Langley's training and educa-tion were extensive. He had gone to an exclusive prep school in Bos-ton. Later he learned engineering on an apprenticeship basis, worked at Harvard, and made original scientific contributions to astronomy through his observations of the sun. Still later, he directed an obser-vatory in Pittsburgh. There he was tapped for an assistant secretary position at the Smithsonian and shortly moved up to general secretary, essentially the directorship. In contrast, the Wrights' education ended before they finished high school, and their work consisted of operating a bicycle shop.

Langley spent his career in large part at the intellectual center of things, Boston and Harvard, an important observatory, the Smithson-ian and Washington. The Wrights' intellectual setting in Dayton was not so rich. When first starting their work, they could not find the books they needed in the Dayton public library; instead, they had to carry on an extensive correspondence to find out what was important and how to obtain it. Langley's resources at the Smithsonian were im-mense. His post was the most important governmental science position of its day. Thousands of dollars and abundant human resources were at his call. The Wrights' livelihood was dependent on work in their bicycle shop. Langley's work style was hands off. He commanded peo-ple to do things and then he stepped back. The Wrights' approach was hands on; every inch of their aircraft had been touched by them. In short, the contrasts are large indeed for these nonspecific background factors.

The problem-specific factors also reveal sharp contrasts. With an occasional exception, Langley did not seek aeronautical principles to constrain aircraft design and construction. He viewed himself as a sci-entist, and his observations and measurements were directed at getting what mathematicians call an existence proof: Manned, controlled, sustained flight was theoretically possible. In contrast, the Wrights wanted to move beyond the possibility of flight to its realization and perfection. To make rapid progress, they tried to rely on the scientific observations of others. For example, they initially accepted Lilienthal's lift tables. But, if necessary, as when the accuracy of the Lilienthal tables became suspect, they were ready to construct wind tunnels and do the basic science required.

Perhaps as a result, Langley possessed far fewer constraints to aid his designs. Constraints help us to navigate through large search spaces of possible design. Here the search space of possible wing shapes and sizes, control structures, and engine configurations was immense, and the Wrights were far ahead in charting a path through it.

Emphasis was also important. Langley's primary focus was on the propulsion component. He spent his time and resources on developing powerful engines with minimal weight. He was well ahead of the Wrights here, because their concerns were with control and lift, and the interrelations between them. In fact, Langley's view was to build from the motor outward toward a complete aircraft. The Wright's view was to build from lift and control toward a motor-driven aircraft.

Langley's major testing device was a large whirling arm that he mounted his models on. It was subject to belt slippage, vibration, and wind gust, to name a few of its problems. The Wrights' major testing device was the wind tunnel. Because a model does not move very much in a wind tunnel, it is possible to measure its performance in many sensitive ways—for example, by connecting strings between parts of the model and by adding a balance scale to measure forces. Measurement with Langley's whirling arm was much more difficult.

The notion of control has already been introduced. Langley advocated automatic control and the Wrights advocated manual control. Langley's view of the wind as a turbulent and unpredictable factor moved him toward automatic control. The Wrights sought out an environment with a steady wind in order to provide reliable lift. Their success in glider flights with their own control system convinced them that manual control could cope with the wind. While constraints are necessary to limit the search space, here Langley imposed an unneeded requirement on himself.

Almost as important as knowing what problem to attack is knowing what problem can be safely ignored. Langley's judgment on this issue turned out be especially faulty; he ignored control and aeronautics in favor of power. The Wrights' approach was completely opposite; they ignored power early on and concentrated on control and aeronautics.

Landing provisions were also ignored by Langley. He launched his flights over water. Both successful and unsuccessful flights ended up in the waters of the Potomac, if lucky. Had a flight gone over land before coming down, the pilot might well have been killed. And a water landing always carried with it the possibility of the pilot drowning. The Wrights carefully considered the problem of landing. Their aircraft had skids and they launched over sand where recovery was easy and the pilot was subjected to minimal risk.

The number of flights made by Langley was very small, in part because launching over water and subsequent recovery were costly in time and effort, even for small models. For scaled-up and manned flights, this was even more so. The Wright's launching over sand made recovery rapid and allowed for many flights each year.

Langley did not use manned gliding, so the flying ability of his pilots was nonexistent when they first launched a powered craft. Langley himself delegated the role of flying, so he had no firsthand experience with its problems. The Wrights were both skilled glider pilots before a motor was added.

An analogy best sums up the contrasting approaches. To Langley the possibility of powered land travel would be explored with the use of a machine somewhat like our modern-day motorcycle running on a narrow track in the dark, a track surrounded by swamps on both sides. For the Wrights, however, the rider would first walk a bicycle with a light on it along the track to assess its properties; then he would straddle the bike while walking along; later he would practice peddling and controlling the bike; then finally a motor would be added only after he was familiar with the track and also the riding and control of the bicycle.

The analogy is not too farfetched. On December 8, 1903, amid much fanfare, Langley launched his motor-driven and manned acrodrome over the Potomac. It promptly crashed, and the pilot was almost drowned. On December 17, 1903, not much more than a week later, on a remote beach in North Carolina, the Wrights produced the first powered, controlled, sustained, and manned flight. No one was hurt, and they proceeded to repeat their feat three more times that day.

Surely, this is an instructive contrast. Langley represents big science at its worst. After spending tens of thousand dollars in probably the most expensive scientific and inventive exploit in history, the result is a stunning failure. The Wrights represent small science at its best. Through intelligence, skill, and persistence they surmounted every obstacle at a minimum of cost. In Crouch's phrase, "a dream of wings" had been realized.

By now, we have come a long way from the woodscrew in my hand. We have turned water uphill with Archimedes, conceptually boarded Leonardo's helicopter, and looked at the twisting wings and propellers of the Wrights. Surprisingly, that screw in my hand is a recent development. Until the last century, it was too difficult and expensive to make screws for a one-time dedicated use as a fastener.

The lessons? A really powerful principle and invention like the screw can turn up almost anywhere. It becomes a building block for an unimaginable range of other inventions. Even as commonplace a device as the woodscrew is not necessarily driven by specific need. Its origins and subsequent applications can range from need-driven, to device-driven, to the deep abstractions of pure mental play, as I have argued in the journey of the screw from Archimedes to Leonardo to the Wrights.

II

AN INVENTION FRAMEWORK

4

Describing an
Invention

Quickly, I stab the crust of the chicken pot pie with my fork, so it won't explode when heated in the microwave. After it cooks, I use the fork to break the shell and stab a piece of meat and transport it from plate to mouth. Then I scoop up some vegetables with the cup or palm of the fork. The crust is too big for the next bite, so I use the edge of the fork to cut it. A piece of potato looks a bit dry and I mash it into the gravy by using the back of the fork. What is going on here? Evidently, something more general than just using a fork. It is almost systematic, almost describable.

How should we describe an invention? To gain a more systematic view of invention, we must develop a framework, a set of categories for describing inventions. If we can describe inventions within this framework by using a roughly common language, then we should be able to group, interrelate, manipulate, permute, add to, delete from, abstract, and generalize on those inventions. We can then compare and contrast inventions; we can classify them and order them in ways that would be difficult or impossible by using the real thing.

Our first step then is to develop a set of categories to use with slight modification across a broad range of inventions. Suitable categories include the *purpose* of an invention, the *physical principle* that underlies it, the *parts, features,* and *dimensions of variation* that we find when we look closely at or inside an invention, and finally we need categories that place the invention in a *context*, in an invention family.

Our descriptive framework for representing inventions is a variation on the *frame,* a data structure used by researchers in the artificial intelligence community to describe complex behaviors and procedures. Lately, it has been used to describe inventions. We are all familiar with the ideas that underlie the notion of the frame. Basically, the frame is like a business or medical form, with a set of standard categories that then take in specific information. Let me illustrate this framework by describing a simple device.

Describing a Fork

Figure 4.1 presents drawings of the table fork and Table 4.1 is the corresponding description. First, we ask what *purpose* or function does this invention serve? As is common for inventions, the fork has several purposes. In Table 4.1 these range from food pickup, transport, and delivery, to sanitation, safety, and manners. An invention's purpose tells us what the invention does at the start of its operation and what it finishes with, like picking up food and depositing it in the mouth. Often, the purpose of an invention is the best place to start a description. In an invention as simple as the fork, numerous purposes are served.

Continuing our descent in the table, another descriptive category is the *principle* by which the invention operates. For the fork, we see that the tines are essentially multiple points or *wedges* that allow for easy penetration into food, the cup or palm of the fork acts as a *container,* and the handle serves as both a *lever* and a *separator,* so the hand is away from hot, oily, or sticky food. So in a tool as simple as the fork, at least four different principles are at work.

It is time to make an important point about the descriptive categories *purpose* and *principle.* We do not directly see these categories. They must be won over to our description by hard conceptual analysis, examination of the historical record, and informed speculation. That analysis is easier in simple inventions like the fork than in complex ones, but the ideas are the same.

In the next division of Table 4.1, we find the category *parts and dimensions of variation.* While an invention's fundamental purpose and principle often change slowly, at the level of parts, the inventor makes frequent changes. Now we are removed from abstractions, and we can actually see movement and action. Palpable things like tines, cup, and handle make their appearance. Corresponding to each part or feature is an associated function, shape, and material. One function of the tines is to spear or stab. Other functions include: scoop, cut, hold down, and poke holes. Each part has a shape that is likely to be specialized for an associated function. Thus the sides of the tines are wedgelike but not as sharp as the edge of a knife. If they were knife-sharp, they

The common eating fork.

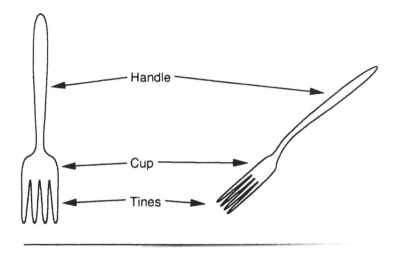

Variations on the common eating fork.

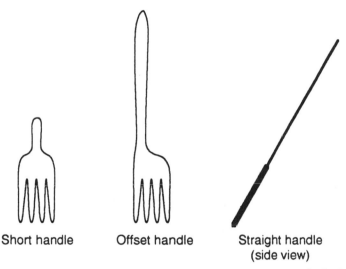

Short handle   Offset handle   Straight handle
(side view)

Figure 4.1 Fork and family relations. The top row contains standard forks; the bottom row some nonstandard variations.

would cut much better but they might cut the mouth also. As a result, their sharpness is a compromise between cutting efficiency and safety.

    Continuing in the table, the category *material* tells us what the fork is made of. Most modern eating forks are made of a metal like stainless

Table 4.1 **A Symbolic Description for the Common Eating Fork**

Purpose: What is the invention used for? What is its purpose? How does that purpose relate to the user? The fork is used to transport food from plate to mouth, with separate phases for the pickup, transit, and delivery of the food. It is also used to avoid touching food with fingers, for reasons of sanitation and to keep from getting burned or messy.

Principle: How does this invention work? What is the underlying physical principle? In the fork, the multiple wedges on the tines are used for penetration, the cup of the fork serves as a container, and the handle acts as a lever for lifting and as separator of hand and food.

Parts or Dimensions of Variation: What are the individual parts and what are their dimensions of potential variation? Parts of the fork are listed in the subtable below along the left side, and dimensions of variation are shown along the top.

			Dimension of Variation (Property or Attribute)		
Part/Feature	Function	Shape	Material	Grip	Movement/Direction
Tines (points)	Spear/stab	Long, pointed	Metal	Precision, power	Down
	Hold down	Long, pointed	Metal	Forefinger	Down, opposite hand
	Poke holes	Long, pointed	Metal	Precision	Down
Tines (back)	Mash soft food	Alternating tines and spaces	Metal and space	Forefinger	Down
Tines (side)	Cut	Dull edge	Metal	Forefinger	Down, lateral
Tines	Scoop	Alternating tines and spaces	Metal	Precision	Lateral

Handle

Human interface
Separation of
hand and food

Long, narrow;
curved

Metal or wool

Varied

Lifting up and
moving to the
mouth

Family Relations: What is the context of the invention? What other inventions are in the same family? Knowing a context will provide a basis for comparing the fork with other inventions, a context from which we can begin to abstract fundamental ideas. Some of these family relations include:

Invention Family: How does the invention fit with other inventions? The fork is an eating utensil.

Ancestors: What is the form of earlier versions of the invention? For the fork, these include the hand and fingers, pointed stick(s), and possibly the knife and spoon, because the fork was late in developing.

Descendants: Later versions of the invention. These include similar forms made of other materials, like the plastic fork (for throw-away that will not have to be washed), and the "spork," a combination spoon and fork that is used in fast-food establishments.

Specializations: What are some closely related inventions? These include the olive fork and the large serving fork.

steel, but sometimes wooden handles are used. And in a certain sense the tines of the fork are made of space as well as metal. Without the space between the tines, it is not possible to spear meat and mash potatoes. Next in the table, we find action features like *grip* and *movement direction,* aspects of use. If the fork is held lightly in balance between thumb and index finger, as in normal eating, the user holds it with a balanced *precision grip.* A precision grip is also used when scooping the food from the plate. If the fork is held with the palm of the hand wrapped around the handle and the thumb is at the top of the handle to vigorously stab something, the user (probably an infant) is employing what we will call a *power grip.* If the forefinger is stretched along the back of the fork's handle, to hold down meat while cutting it with a knife in the opposite hand, the user is employing a *forefinger grip.* The fork's direction of movement also varies with the function and grip. Variations in grip and movement direction work together to make possible many different functions for the same physical fork.

To understand an invention, it is useful to begin with a table, as we have done. Even highly technical inventions can be understood by the interested general observer at these levels of description. For the implementation of an inventive idea, however, the details of relationship between part and function are often crucial. To see these details in the fork, we introduce a *test for design.* The basic idea is that by varying a component, we will be able to grasp how essential it is; we will bring out the hidden intelligence behind it. In the bottom part of Figure 4.1, the relations among the parts of a fork were altered to show that something more than a listing of features is required to capture an invention.

Why is it that the fork did not take one of the alternative forms shown in Figure 4.1? The short-handled fork will not work as well as a normal fork because it is hard to grip. The offset fork is unbalanced. The straight fork does not allow for effective scooping and pickup; we cannot load a straight fork with our peas and prevent them from running downhill. Nor does the straight fork allow for effective transit of those peas because our normal precision grip on the fork handle is one in which the handle points downhill (try it). This same downhill angle also affects food delivery at the mouth; we want the food to enter the mouth on a horizontal or level plane. All these transport requirements for pickup, transit, and delivery are realized by having the fork's cup dip down from the handle and then angle upward slightly. The present arrangement of the fork's components is a triumph of functionality over many bad designs that might have been, another example of the hidden intelligence of everyday things.

Continuing with Table 4.1, we find the descriptive category *family relations.* This category establishes a context for the invention. Essentially, family relations are different forms of superordinate categories into which a fork may be placed. In a functional sense, a fork belongs to an invention family that we may call "eating utensils." Other family relations include ancestors, descendents, and specializations. *Ancestors* are the more or less distant progenitors of an invention, often progenitors that bear a resemblance in purpose. Because the historical record is fragmentary, we must often make guesses as to ancestry. Those guesses serve to lend structure and possibility. Most certainly, hands and fingers are distant ancestors of the fork, and the knife and spoon are closer ancestors. Interestingly, the early forks of the eleventh century often had only two tines. This suggests an invention path in which a generalization is made by starting with a pointed stick or a knife and then adding more points. But another origin, one based on shape rather than function, is also possible. Pitchforks have existed from Neolithic times, so the fork may be the result of scaling down a pitchfork. Undoubtedly, the actual origin is not recoverable; nonetheless, speculation about origin helps us create a family context for the fork.

To continue with the idea of family membership, *descendents* are later versions of an invention. The plastic throwaway fork is one example and the combination spoon and fork, termed a spork, is another. A descendent is not necessarily an improvement. Finally, *specializations* are a kind of descendent adapted to a particular environmental niche. Examples are forms like an olive fork or a large serving fork.

Even if we have a complete and accurate description of a fork, that is not enough to master the tool, as any infant's use of a fork or spoon confirms. Effective use of a tool requires a procedure for using it, a sequence of actions to accomplish a given function that we will call a *use script*. A first attempt at stating a use script is:

Use   = Part or Edge   + Grip   + Action

For the fork, particular embodiments of that abstract script can be described this way:

Stab	= Tine tip	+ Precision grip	+ Downward thrust
Scoop	= Tine tip	= Precision grip	+ Push sideways
Cut	= Tine side	+ Forefinger grip	+ Downward, lateral

Thus for the fork's stab function, we use the tine tips, a precision grip, and a downward action.

This is not an arbitrary script, as seen by mixing things up in a variation of the test for design. Using a precision grip to cut a piece of flat lettuce with the side of the tines just does not work. Nor does stabbing work when we use a scooping action. Each use or function involves a careful selection among part, grip, and action.

The fork is more complicated than it first appears. As expected of a tool with multiple working edges, grips, and associated actions, its parts have many functions and the overall tool has many purposes. Its complexity was revealed through our descriptive framework. The idea of describing inventions is a useful one. Some categories approach universal applicability across inventions: purpose, principle, parts, dimensions of variation, and family relations. Yet different families of inventions require other categories to capture the necessary detail. For example, grip is relevant to tools but not to a business form. Other categories such as name, address, and age are needed for the business form but not for the tool. Any good descriptive framework involves regular tradeoffs between generality and completeness of its categories; but to describe an invention, we should begin by being greedy, seeking both generality and completeness of detail wherever possible.

Finding Parts and Dimensions of Variation

The descriptive categories of purpose, principle, and family relations are readily understood, even if extracting them from a given invention is not always easy. However, the complex category called *parts and dimensions of variation* requires more elaboration. Indeed, to construct a satisfactory description, it is often necessary to move back and forth between the evolving description and two other activities called *parsing* and *identifying dimensions of variation*. To see the generality here, we need to go beyond the fork.

Parsing, breaking things down into meaningful units, is an important and inadequately understood strategy. For our purposes, it is taking an invention apart, mentally or physically, to understand and work with it. Parsing is often the first step in identifying functional parts. When parts are identified, then the dimensions of variation—the aspects that can be changed or varied—begin to pop out. Of course, this is not necessarily a sequential thing. Often the identification of parts and dimensions of variation are simultaneously pinpointed, and several cycles of alternating between the identification of parts and dimensions take place. The importance of these processes is underscored by the inventors I have talked to. Frequently, an inventor will claim to get ideas by taking apart an existing invention so he or she can find parts to change for greater efficiency or enhanced performance. As noted before, the Wright brothers were notably successful in parsing the problem of flight into three subproblems (lift, control, and power) so they could attack them separately.

Parsing can take place on several levels. It can focus on an invention's perceptual form, its function, the interface in making contact with it, the actions in using it, the grip used, and the materials it is composed of. In short, any descriptive category can be a basis for parsing. Let's consider each of these.

Perhaps the easiest approach to parsing, or dividing an invention into its parts, is to look for perceptual discontinuities. On a knife, for example, the point, the cutting edge, and the handle are often perceptually discontinuous parts of the knife, because they have different textures or materials, or sharp boundaries between them. When an isolable perceptual region is discovered, it is reasonable to inquire about corresponding differences in functionality. Certainly, the perceptual entities *point, edge,* and *handle* have distinguishable functions. In a more complex inventions, rates of motion may also point to isolable parts. In black-box inventions—where components are not transparent, as with electronic or chemical inventions—parsing is particularly difficult, and it may require elaborate instrumentation and theory to disentangle the separate parts. A simple tool without moving parts, like the fork or the knife, is the easiest to parse.

Instead of beginning with perceptually distinguishable features, we can turn the process around and begin with distinguishable functions as the basis for parsing. For example, with the knife such functions would include cutting, making holes, and scrapping. Each of these functions is likely to emphasize some parts rather than others. A consideration of functions can lead us to parts, and vice versa.

Any useful human artifact will have an interface for people to interact with it. Sometimes that mode of interaction takes a long time to perfect. For example, the earliest stone tools did not have an explicit handle as an interface at all. At a later date we see the appearance of an explicit handle. The development of better handles and interfaces is a common invention path. To understand an invention, we must understand the interface between human and artifact. The mark of a successful invention may be in the human interface it provides.

Closely related to the level of function and interface is the level of action. How is the invention used, which edge or portion of it is active and in what way? For example, depending on which way a screwdriver is turned, a screw is either inserted or removed; these are inverse functions of one another defined by the direction of rotation. The same part, the screwdriver tip, has at least two functions, insertion and removal of screws, that are determined by the associated actions.

Even though the same tool edge and action are used, different grips may result in further specialization of function. Whether a power grip or a precision grip is used, this gives rise to different capabilities when drawing a knife across a piece of work. A power grip is with the whole hand wrapped around, like that on a baseball bat; it yields maximum force and minimum control, providing deep and rough cuts. A precision grip is with the thumb and forefinger acting in concert, like that used with a pencil; it supplies maximum control and minimum force, allowing small-scale accurate cuts.

Different materials in an invention or tool often define different functions. Thus the handle and working edge of a tool are usually recognizable by virtue of the materials of composition. For a knife, the handle may be wood and the blade steel. This example is seemingly trivial because we know from experience the difference between the blade and the handle of a knife, but when we are looking at an unfamiliar invention, a parse by materials can be helpful in establishing the different functions of the parts.

Of course, parsing is not an end in itself. Once an invention is parsed, we can begin to assign to it a description consisting of separate categories, like parts, functions, materials, and interface.

Concurrent with parsing and the development of a description is the identification of dimensions of variation—those things that can be varied. Knowing what can be varied is an important step in knowing what to do next, in establishing a "space" of possible variations on an invention. Changing components may lead to an improved invention, to a new invention altogether, or to a disaster. Knowing what can vary and how to vary it is one of the requisites to expert invention and design. Almost any of the parsing categories that we have just identified can serve as a dimension of variation: form, function, action, and so on. But there are still other ways of identifying dimensions of variation.

One way is to find possible dimensions from other related artifacts or procedures. For example, it is obvious that the muscles of the back are fixed in location and function. Yet surgeons routinely move body parts around for reconstructive purposes. It should not come as a surprise, therefore, that a new experimental technique moves a back muscle around a damaged heart, keeping vascular and nerve connections intact. The back muscle can then be stimulated with a pacemaker to assist the injured heart pump blood until it can recover. Evidently, the location of a body part like a muscle is seen as a dimension of variation by a surgeon.

A second way to spot dimensions of variation is to let the eye be moved by the flow of history. Comparing the same invention at two or more points in development often reveals otherwise latent dimensions of variation. It took me an embarrassingly long time to realize this. I stared at different stone tools over many months before I realized that an explicit handle took many thousands of years to develop. Then it became clear: an interface is a potential locus of variation; it can be changed.

Still another way to discover forms of variation is to create them. This may be done by applying some standard principles, like looking for an inverse of an action. For example, if a nail can be pounded in, there ought to be a tool and a way of removing it. Most hammers in fact have this capacity.

These methods of finding dimensions of variation can be used systematically. Collectively, they constitute a different way of looking at the world: Instead of seeing the world as it is, we begin to see it as it might be. In short, reality is simply one of many possible worlds. Once the seeing of possible worlds becomes a habitual perception, the next step is to search for evaluative characteristics of those worlds: What distinguishes one possibility from another as better or worse? Ideally, we may even discover an optimal world, one that is the best of its kind. That best world may be exceedingly complex, or it may be a microcosm as simple as a well-crafted knife or a fork or a hammer handle that has been carefully shaped to provide the perfect fit between task and hand.

5

Evaluating and
Comparing Inventions

A mental lightning storm is going on in my head. I'm acting as the psychological coach for the chess master, and he keeps ignoring my advice. I want him to consider more candidate moves before making an actual move, and he fights my suggestions and insists on considering only three or four moves, carefully evaluating each one in turn. He can make scores of moves but persists in only considering a few. He shouldn't be evaluating his moves until he has generated a number of candidates. Why won't he listen to his brainstorming coach?

Brainstorming as a method tells us that we should put off evaluation until we have generated a number of ideas. In certain contexts it can be quite useful. However, contrary to the advice of brainstorming, a useful skill that promotes invention, as it does chess, is the ability to evaluate quickly, realizing that we may need to backtrack if our chosen move is unsatisfactory. Similarly, in invention a rapid evaluation identifies those aspects that are wanting and can then become candidates for further invention and improvement. For any given invention, the potential forms of evaluation usually are numerous. Let's look at some fairly general ones, without any necessary order of importance.

First, we can evaluate the *overall invention*. As a package, how well does the invention work? Does it do what we want in an efficient and economical way? Alternatively, we may look at the opposite side of the coin and ask, what is wrong with the invention?

Second, we can move away from the whole invention and get

more analytic. What *parts* do not work well? An invention may work quite well overall, but still have parts that can be improved. If we can identify parts that do not work, then they can be experimented with and manipulated much more readily than the overall invention.

Third, we can look for *historical dimensions* of evaluation. What trends drive the development of the invention, from early forms to its present state? Knowing those trends may guide us in seeking the next steps in the development of the invention. This is particularly helpful when we don't know what evaluative standards to use.

Fourth, instead of comparing the target invention with its ancestors, we can make comparisons with a present-day *related invention.* For a given function, how well does our target invention fare with respect to its relatives? A thorough comparison of related inventions may suggest borrowing parts or features from one invention to use in the other.

Some Examples of Evaluation and Comparison

Evaluation and comparison, we noted, go hand in hand. This is because evaluation on an absolute basis is difficult. Having a comparison makes evaluation easier. To illustrate this idea, I want to return to the humble fork.

Table 5.1 shows the main functions of the fork. We see that for spearing foods at dinner it is excellent in effectiveness. It is better than a knife because it is safer. Offhand, for spearing I cannot think of anything better than the fork. For scooping, it is at least very good. Clearly, it's better than a knife because it holds more and does so more readily. Peas are simple for a fork but devilishly difficult for a knife. But the fork is certainly worse for those peas than a spoon, because the spoon holds more in quantity and in variety. Similarly, the spoon will hold liquids but the fork will not. For cutting, the fork is only good at best, but it is better than a spoon, because the fork has a longer and straighter side edge. Still the fork cuts worse than a knife because the knife's edge is longer and sharper. Finally, for perforating many materials the fork is excellent. For many eating and cooking tasks it is better than either a spoon or a knife. Imagine prior to putting a pie in the oven that you poke air holes in the crust with anything other than a fork.

Other dimensions of evaluation are possible. For example, we could have looked at cost, ease of use, manufacturing ease, and no doubt several other categories as well. These evaluative criteria are more or less standard in that they apply to any invention. Others might be more specific, like how easy it is to clean forks relative to knives.

Now let's look at a less obvious example of evaluation and com-

Table 5.1 A Comparison of the Fork and Its Family Relations

Functions of the Fork	Overall Effectiveness	Evaluation	
		Better than a:	Worse than a:
Spear/stab	Excellent	Knife because it is safer	?
Scoop	Very good	Knife because it holds more	Spoon because it does not hold as much and will not work for liquids
Cut	Good	Spoon because it has a longer and flatter cutting edge	Knife because its cutting edge is shorter and duller
Perforate	Excellent	Spoon because it has sharp and penetrating points	?

parison, the lowly nail. We ask the question: "What does a nail do well or poorly?"

Historically, the nail is a very important invention because it allows for the easy construction of complex forms by fastening together simpler components. Pieces of wood can be combined into a bench instead of carving it out of a log. The nail also offers advantages over ancestral methods of fastening such as lashing together, sewing, or using wooden pegs. These other methods are more time consuming because they require separate operations like making holes. The nail makes its own hole and simultaneously acts as a fastener.

I became aware of the nail's advantages as a fastener through a circuitous path. When I start to write about an invention, I often begin by writing down, over a period of time, all the characteristics of the invention that I can think of. Table 5.2 shows some things a nail does well or poorly, with Pro " + " and Con " − " beside each item. Most of the pros and cons of the nail are evident and require no comment. But I must say that I was surprised at finding the sheer number of evaluative characteristics that are related to the nail.

During my evaluative analysis of the nail, I made an accidental discovery that provides a useful aside. When I write, my first draft is usually in handwriting—for superstitious reasons, a yellow pad works much better for me than a white one. Then I might do a preliminary edit of the draft, also in handwriting. At that stage I enter it into my

Table 5.2   Evaluative Comparisons for the Nail and the Screw

Nail	Screw
+ Easy for most people to insert*	− Hard for many people to insert
+ Goes in fast	− Goes in slowly
+ Resists shearing forces	+ Resists shearing forces
− Only moderately good at resisting pulling forces*	+ Very good at resisting pulling forces
− Removable, but surface may be marred by the hammer claw in the process*	+ Removable, and surface is not likely to be damaged
− Not readily reusable after removal	+ Readily reusable after removal
− Unsuitable for some materials, such as metal*	+ Suitable for many materials, metal, plastic
− Rusts	− Rusts, if steel or iron
+ Easy to manufacture*	− Difficult to manufacture
+ Cheap	− Relatively expensive

Note: Pro = +; Con = −.
* = items suggested by comparing the screw with the nail.

computer, where several drafts later the final work emerges. The hand-writing part is often a process of slow accretion: a few notes on a given topic one day, more the next, and so on. Although yellow pads are important for me, I don't feel the same fidelity toward pens and pencils; a variety were used, which turned out to be important. After a preliminary listing of the pros and cons of nails as fasteners, I ran out of ideas. Another day I started with a different-colored pen and could not think of anything to write about nails. Then, for some reason, I started another list with pros and cons for screws. As usual, the first ideas came easily—and, surprise, they suggested additional pros and cons for nails as well. The different colors of ink provided a dated trail and revealed an interesting truth: Writing about screws suggested additional evaluative aspects for nails. In Table 5.2, those items about nails that were later suggested to me by screws are starred. For example, I had not originally thought of nails as easy to insert until I had listed screws as difficult to insert.

To generalize on this idea: Thinking of evaluative dimensions or criteria is easier to do if there is a *comparison invention* in the same family to also examine, a *B* to contrast with an *A*. Each will reciprocally feed ideas into the other.

After identifying evaluative criteria, we need to apply them. Criteria are often applied incrementally, to produce a fine-tuning of a part or of the relation between two or more parts. As much as possible, good things are maximized and problems are minimized. But with evaluation and incremental improvement it is easy to get stuck in a local optimum, something that looks good from a limited perspective. An example is an evaluation that is particular to an invention form: One could easily spend a lifetime perfecting a longbow and miss the idea that gunpowder-based arms might be superior. Or that flapping wings, however well perfected, will not well serve human flight. Or that a knife based on the wedge principle is one way of cutting, but a laser beam can also cut, even though the principle of cutting is much different and may be superior for fine-grained surgery.

These last examples suggest that in contrast to incremental improvement, a radical evaluative action is to conclude that one is not operating on the right principle. This may mean finding a new principle of operation, like moving away from the longbow to discover gunpowder and firearms or away from flapping wings to discover fixed wings with lift. Even if the invention path is toward a new principle, many evaluative criteria may carry over intact from an old principle. The transition from longbow to gunpowder again provides an example. Long before the advent of firearms, evaluative criteria like accuracy, rate of firing, penetration, and range of effectiveness were used in the design of longbows. So when firearms were introduced, people were able to use many of the same evaluative criteria. People already knew what they wanted a firearm to do, even if they did not know how to go about getting it. Probably, the development of firearms was greatly speeded as a result.

I used to believe that evaluative criteria were specific to particular domains of thought and invention. In support of this contention I've sometimes found highly specific evaluative criteria suggested. This is one from the imaginative mind of D. Lenat, a worker in the field of artificial intelligence: In number theory, if a number has many factors (like the number 60 which has as factors 2, 3, 4, 5, 6, . . .), it is interesting; if a number has very few factors (like the number 3, which only has itself as a factor), it is interesting. These two criteria of interestingness are highly specific evaluative criteria without general applicability beyond numbers.

Yet, many evaluative criteria are much less specific. In fact, we may distinguish evaluative standards with middle-level generality from those with high-level generality. Examples of middle-level generality include the criteria in place for the longbow before the advent of firearms: rate of firing, range of arrow, accuracy, and penetration into target. Each of these should be maximized, and each transferred directly to firearms. However, these same criteria are not suitable for hand tools, or a variety of other domains. Hence they are of middle-level generality.

At a higher level of generality are evaluative criteria that cut across the broad domain of firearms, tools, and many other things. They include:

- *Workability*. Does the invention successfully do what it is supposed to do?
- *Portability*. Reductions in size or weight are frequent evolutionary trends in invention, particularly in firearms and tools.
- *Ease of use*. Changes in a tool are often in the direction of improving the human interface, as indicated by the development of the knife handle.
- *Durability*. How long a tool lasts is likely to be inversely related to its ease of manufacture. Movement to more lasting materials often occurs, say, from wood to stone to bronze to iron to steel, as artisans become more proficient in the technologies of harder materials.

How can evaluative criteria, from specific to general, facilitate invention? Several ways. First, the criteria tell us the direction in which to move. They constrain our search for new inventions, or for variations on existing inventions. Some evaluative criteria need to be maximized (portability, durability) and others to be minimized (breakage, cost). By knowing in which direction to move, we may be able to borrow or import wholesale the evaluative criteria from another related invention, like moving from the longbow to a firearm. Second, the different evaluative criteria may be played against one another to further constrain and direct search, as in the comparison of the nail and the screw.

Many criteria will be compatible with one another; we can have portability and range of use at the same time. Other criteria will be incompatible; we cannot have the features and workmanship of a Mercedes at the price of a Chevy. Still other criteria will have an ambiguous relationship to one another: How do difficulty of manufacture and ease of use interact?

This leads us to an important trick underlying invention and design: Incorporate in the new invention the compatible criteria of related inventions, and judiciously trade off the incompatible criteria against one another. While the trick is easily seen, the implementation of it requires great slight of hand and insight, particularly in achieving the right balance in the tradeoffs among incompatible criteria. With the appropriate design, materials, and manufacturing sophistication, that Chevy can take on a number of the Mercedes' characteristics.

Now that we have new concepts like parsing, dimensions of variation, and evaluation behind us, we can look more closely at some real examples to see how our descriptive framework comes together to capture the essence of some real-world examples.

6

Understanding the Created World

I'm looking at the soup ladle in Figure 6.1. Why is it shaped this way instead of like the serving spoon next to it? What is the purpose behind its distinctive shape?

Finding Purpose

Often we begin our understanding of an invention by determining its purpose. Strangely, it's not always clear how to do this. Let's look at two simple inventions, the soup ladle and the door knob, to see how we can determine purpose. Along the way, we will find out if it is a bad thing to be as dumb as a doorknob.

To appreciate the soup ladle, we capture its special functionality in comparison to that of an ordinary general-purpose serving spoon. In other words, we are explicitly making use of a comparison with a closely related member of the same invention family. Figure 6.1 shows the two kinds of spoon. The shapes differ. The bowl or cup of the ladle is bent up and the end of the handle is bent back around. Also the cup of the ladle is larger than that of the serving spoon, and the handle is longer too.

Why these structural differences between spoons? What is their purpose? Perhaps they have to do with the food and the container from which it is extracted? The regular serving spoon is commonly used for solid foods in relatively shallow large dishes; its shape is well

Serving Spoon   Soup Ladle

Figure 6.1 Serving spoon and soup ladle, two members of the spoon family, with quite different purposes.

adapted to the task. But if used to serve soup from a deep container, we would find it difficult to get the soup up without spilling it out of the spoon's cup. So we can see the reason for bending up the cup of the soup ladle, relative to the handle. The purpose of the bent up cup is to provide a way of entering and exiting deep containers without losing the soup from the cup of the ladle. The soup is also likely to be hot, so the long handle will help protect the hand and avoid getting burned. Finally, we come to the end of the ladle handle that is turned back on itself. To see why, suppose that it were straight. If the straight handled ladle were left in a big soup pot, it might slide and fall into the soup.

How did the structure of the soup ladle come to pass? I don't believe that any one person sat down and said that a soup ladle will have these three purposes. Instead, it is likely that the features of the soup ladle gradually evolved to handle each successive problem or annoyance. Until recent times, that is the way of most invention, the slow accumulation of intelligence applied to making an implement better and better.

Now let us turn the process around. What if we tried to use the soup ladle for the common uses of the general-purpose serving spoon? Suppose that we are serving peas from a shallow serving dish. The ladle's long handle gets in the way; its rounded, bent-up cup is at the wrong angle for a pushing motion to get under the peas; and the turn-back at the end of the handle is useless.

In each case, the different purposes of the soup ladle and the ordinary serving spoon become clear as we interchange their normal uses and features. In essence, we have executed a variation on the test-for-design by using two different spoons, each for the other's normal functions. This suggests a comparison principle: To identify underlying purposes, try exchanging the features of related inventions. Often the resulting changes in functionality will reveal those purposes.

But what of something so simple as a doorknob. Most of us have heard the expression "Dumb as a doorknob," and it made me wonder about the intelligence of this ubiquitous hardware. Before giving my assessment, let's look at some questions. Why is a doorknob the size and place it is? We can determine its purpose by arranging a series of tests-for-design.

First, the size. The left side of Figure 6.2 shows a door and a number of different sized knobs on it. Ignore their position—we will consider that later. Examine the size issue by considering some limiting cases. The top left corner is a very small doorknob. In the limiting case it is the size of the shaft that the knob turns to activate the latching mechanism. Such a small knob will not give much leverage or mechanical advantage. It needs to have a bigger diameter to provide more leverage. The leverage limitation of the small knob is due to physics. The knob is comprised of a wheel and fixed axle and is therefore a variation on the lever; small knobs provide little leverage and require large effort to turn. The bottom right corner indicates a huge knob; it provides lots of leverage. Perhaps the large knob should be used? But a problem presents itself. The knob is too big to get a single hand around. The limitation of the large knob is a human interfacing problem.

So we are driven to compromise; practical knobs must be halfway between the constraint of physical leverage and the constraint of the human interface limited by body structures and size. The purpose of the normal sized knob is to effectively turn the latching mechanism with just that compromise in mind—neither too small for leverage nor too big for a single hand.

Now for the placement of the knob on the door. Figure 6.2, on the right, shows a variety of knob locations. Why should we choose one over the other? Where should the knob go? At least two considerations will drive our choice, the simplicity of the linkage between knob and latching mechanisms and the leverage required to open the door. First, the linkage. The latching structures connect with the knob at one end and with a recess in the door frame at the other. For economy

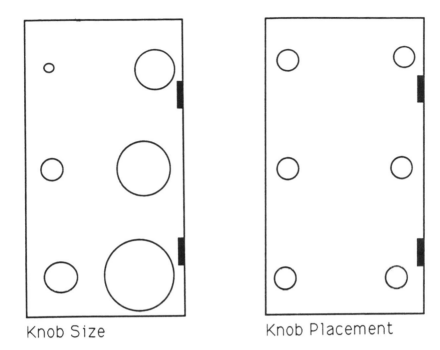

Knob Size   Knob Placement

Figure 6.2 The hidden intelligence of doorknobs, as revealed by varying their size and location.

and simplicity, this linkage should be as short as possible. That means the doorknob should be close to the edge of the door, opposite the hinge side.

Next the leverage. Human considerations might override the decision to put the knob on the edge. Such considerations include leverage. Any lever system has as its components effort, load, and fulcrum. The fulcrum for the door is the hinge, the load is the weight of the door located approximately at its center, and the effort is the force we exert in opening or closing the door. The maximum leverage or mechanical advantage will occur when the force we exert is located far from the fulcrum and load; it should be on the side of the door opposite the hinge. If this is not clear, think of a wheel barrow, with the wheel corresponding to the hinge of the door and the handles as the knob.

So two very different considerations—simplicity of mechanical linkage and adequate leverage—have both led to the same conclusion: Locate the knob on the far side opposite the hinge. Neither of these considerations tells us how high to put the knob. According to the criteria considered so far, the knob could be at any height along the edge opposite the hinge. Once more a human interface problem arises. If we are going to be standing when we open the door, and if we open it with a hand, then the knob should be at arm's height. In fact, that is where we find it.

The knob's hidden purposes include its placement so the mechanical linkage is simple, the leverage is good, and the human fit or interface convenient. These multiple purposes are not obvious when we casually inspect doorknobs. For most purposes the standard placement of the doorknob is best, another example of hidden intelligence at work. Doorknobs are far from dumb.

Finding Principle

It is very cold outside. Around zero, with a wind blowing hard to bring the windchill down to about thirty below. Fortunately, I have a new winter coat-parka that is perfect protection for such weather. It is illustrated in Figure 6.3.

Just as finding purposes is important, so is finding principles. Once again, this is not always easy. However, my winter parka will guide us. The parka has several fastening mechanisms. These are based on different fastening principles, each one appropriate for its particular task. Let's look at them.

The closure of the parka hood around the neck uses a drawstring that is tied. The tie is certainly one of the oldest fastening principles, and one of the simplest. This one has a few refinements added. It is enclosed in a sewn channel so it will remain in exactly the right position when the drawstring is pulled tight. The string itself has some wooden balls at the end so it will not pull inside the channel. The string tie principle is cheap and adjustable, but it does take two hands and some dexterity to use it.

Surely, the drawstring is enough for the hood? No, in addition there is a flap that pulls across the throat for added protection from the wind. The principle of securing the flap is that of a snap with a flanged cylinder or knob that fits into a receptacle with a spring. The flange opens the spring which then locks about it. The spring is open on an edge, however, so the snap can be disengaged by pulling hard on the two surfaces and spreading apart the spring. For the user it is possible to close or open the snap with one hand, after a bit of practice, but it's hard to do if you have a glove or mitten on.

The front of the parka is closed with a strong zipper. The principle of the zipper is a set of discrete interlocking clasps that are brought together by a continuously moving slider. When unzipping, the other end of the slider wedges apart the little clasps. My parka's zipper has two different sliders, so I can open or close it from either the top or

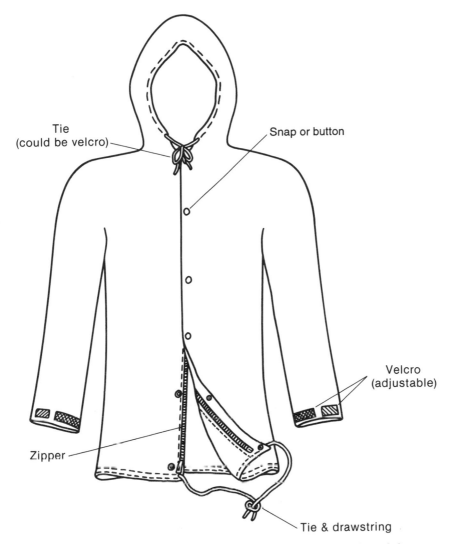

Figure 6.3 The principles of fastening revealed in the parka. The subtle nature of different fastening principles is indicated by trying to swap them around. It is hard to find an improvement, but easy to find a bad combination—like a tie scheme for the cuffs.

the bottom. That flexibility allows me to adjust the air flow and temperature as needed. To get the zipper started requires two hands, but adjustments of it are readily made with one hand, and in the dark. This zipper is not made of metal but of a hard plastic. Assuming that the plastic is strong enough to hold up, it's a good material for the user because metal zippers are notoriously cold.

There is still more to the zipper. It is covered by a flap that is secured by a row of snaps. The flap helps to keep out drafts but I suspect that the main reason is to protect the zipper from mud and dirt, substances that zippers do not appreciate.

The ends of the parka sleeves can also be adjusted to keep out drafts. The fastening principle here is Velcro. The way Velcro works is to bring two different surfaces together, where one surface is a set of miniature hooks and the other a set of miniature eyes. Under contact, the two surfaces "grab" one another. Because both the hooks and eyes are flexible, Velcro surfaces can be unfastened by pulling up to straighten the hooks. Velcro is a very good fastener for sleeve ends because it requires only one hand to work. Older fasteners like snaps did not allow for continuously variable tightness, and they were awkward to work with one hand.

Each pocket on my coat has a flap on it, and the flap is also secured by Velcro. This makes closing the pockets a fast, one-handed operation that can be done in the dark. I have seen some coats with zippered pockets but that is probably a style matter. Other coats use snaps on the pockets, but snaps are slower and less convenient than Velcro.

All of the fasteners described so far are reversible ones, under convenient control of the user. But the parka has other fastening principles at work also. The separate parts of the snaps are fixed to the fabric by a riveting principle, two interlocking pieces of metal fit through a hole in the fabric. Under pressure they are squeezed together in a permanent way. The plastic teeth of the zipper are attached to the zipper backing in a way I do not fully understand. My guess is that a thermoplastic process was used here with hot malleable plastic squeezed down on the fabric backing. Finally, one of the oldest forms of fastening is used to hold all the parts together. It is called sewing.

A good trick for understanding principles is now revealed. Look at an invention family like fasteners and note the structural and functional differences among the members of the family. The interrelation of function and principles is delicate. I doubt that any interchange of fasteners and locations on my coat would improve on its functioning. I think it is a remarkable fact that a winter parka uses so many different fastener principles to get its job done.

Finding Parts: Structure, Function, Material

It's time to look more closely at the detail. Now for the first time we are talking about palpable perceptual forms: the structure, function, and material of parts. To set the stage, let's begin with a quote from Robert Fulton, the inventor of the first practical steamboat, someone who has much to say about the importance of parts:

> As the component parts of all new machines may be said to be old . . . the mechanic should sit down among levers, screws, wedges, wheels, etc. like a poet among the letters of the alphabet, considering them as the exhibition of his thoughts; in which a new arrangement transmits a new idea to the world.

Well put indeed, so let's begin with the structure and function of parts. The first and simplest case to consider is the single part with a single function. A simple example is a wooden peg for hanging a coat. It juts out from the wall and the coat is draped over it. Actually, examples this simple are hard to come by; the norm seems to be that a given part usually has several functions.

The second case to consider is the part with a single structure and multiple functions. The opening in the top of a cup is a good example; it is used for both filling and emptying. By virtue of time-sharing, like a condo in a resort, the top at one time has fluid poured in and at another has fluid poured out. A closely related variant is a part that appears more than once, and the same function is duplicated many times. The bricks of a wall serve as an example, but even here not all bricks are strictly equal. In practice some bricks are used for structural purposes and others for ornamentation, some for walls and others for arches. The multiple wheels of a cart provide an even more complex variant. Each wheel provides support and a relatively frictionless rolling action. However, an emergent function arises when three or more wheels are used; then the cart is stable and does not tip at rest. Notice that stability is a property that no one wheel contributes on its own.

The third case is even more interesting. Consider when a part with multiple functions splits in order to provide a separate structure for each function. We have already touched upon this with Carelman's coffeepot. Unlike the cup with a single orifice for filling and pouring, the coffeepot has two separate and specialized orifices: one for filling, the other for pouring. By separating the functions into distinct parts, each function can then be fine-tuned and specialized independent of the other. This is a common idea, and it means that time-sharing is no longer required. The double swinging doors in a restaurant for entering and leaving the kitchen provide another example. By the duplication and separation of doors, conflicts and accidents in entering and leaving the kitchen are avoided.

The fourth and final case is just the opposite of separating function into different parts. It is the integration of previously separate functions in one structure or part. An example is the Swiss Army Knife blade that is both a can opener and a screwdriver. Another is the Wright brothers' use of a wing for both lift and control. The integration of functions in the same part is a common manufacturing strategy. The auto manufacturer tries to reduce the parts count by combining numerous functions on a single part; frequently the integrated part can be stamped all at once and then assembled in one swift motion. A triple economy is affected, in parts, materials, and assembly.

But isn't this all contradictory? Sometimes separating parts and sometimes combining them. What is the logic here when so many contradictions abound? It is a logic not of consistency and truth, but of limiting cases and practicality.

Yet another way to look at parts is from the view of materials and their functions. An ordinary pencil consists of a wooden shaft, a graphite core, a rubber eraser, and a crimped metal cylinder that holds the eraser and fastens it to the wooden shaft of the pencil. All straightforward enough, but the obviousness of the example conceals the intelligence behind the particular choices. Certainly we will not use the metal for the eraser and the graphite for the holder. All the parts are made of an appropriate material, and each material is in exactly the right place to ensure correct functioning. A substitution or swapping of materials will impair the functionality of the pencil. In invention, as in evolution, random changes are likely to be lethal.

The role of materials in invention is underscored by our working example of the fork. Some forks have parts made of different materials, a metal working edge and a wooden or plastic handle. For simplicity, let's suppose that the fork is all one big part made of the same material. A wooden fork is fine for serving salads but it may not clean as well as we like. And putting it in the dishwasher will damage its finish. A throwaway plastic fork makes sense for picnics and fast-food establishments. It is cheap enough not to bother with for cleaning but it frequently breaks, it is aesthetically unpleasing, and it is not recyclable and therefore an environmental hazard. An iron fork is strong and durable, but it rusts. The stainless steel fork is also strong and durable, but in addition it is rust resistant and easy to clean. However, it is also relatively expensive and would not be best for picnics or fast-food restaurants. Materials make a difference. They define an artifact by virtue of their properties. Like Robert Fulton and his vocabulary of shapes, inventive minds also have a vocabulary of materials to draw on. Once we think about it, the importance of materials is obvious. But we usually don't think about it.

Finding an Invention Family

Another descriptive category, the *invention family*, provides a powerful step up in abstraction beyond the palpable reality of a concrete invention. It places an invention in a context. That context is invaluable for suggesting generalizations or extensions of the invention or for providing connections and links to related inventions. The family description is likely to influence our thinking about new uses, and some family contexts may be better than others in their suggestiveness. For example, the fork described before can be a member of many invention families:

- *An eating implement.* The context then includes knives and spoons.
- *A hand tool.* The context becomes saws, razors, screwdrivers, and so on.
- *A tool.* Now any tool becomes a member of the same context.
- *An artifact.* Here the fork takes up occupancy with any part of the made world.
- *A piece of metal.* The context now includes airplanes, computers, bulldozers, cars, and pennies.
- *A silver-colored object.* The context includes airplanes, aluminum siding, and polished metal in general.
- *A pointed thing.* Now the context encompasses shapes like knives, swords, bayonets, and pointed sticks.
- *A tool with three or four prongs.* This includes pitchforks and some garden implements.

Which of these superordinate categories provides the correct family context? That is the wrong question to ask, because all of them are "correct" from some point of view. If not "best," are some invention families, some superordinate categories and contexts "better" than others? Probably, but it depends on our purpose. For most purposes saying that a fork is a piece of metal is not very informative. But if our purpose is to look for uses of metal, then it makes sense. A preliminary try at defining the "best" invention family is to claim that its context is composed of a meaningful family of members, that the relations among its family members are more numerous than for a less appropriate family, that the members of the family are somehow close to one another in the abstract space of invention. These common relations may then suggest potential gaps, extrapolations, or next steps in that invention space— in short, possible new inventions.

Why should some invention families provide a better context than others? The answer is a joint one: partly because of the similarity of things and functions, and partly because of the way our memories group together things and functions. A clearer understanding comes from the work of Eleanor Rosch, a psychologist who has worked extensively on natural categories. Rosch asked people to fill in sentences like:

< A _____ is a bird. >

Many people fill in entries like "robin" or "sparrow," but virtually no one fills in "penguin" or "ostrich." The reason for this, according to Rosch, is that our memories are so organized as to emphasize prototypes of categories, typical members of the category. Of course, if we lived in a different place in the world, penguins or ostriches might well become our prototypes.

How do Rosch's results relate to finding prototypical or best invention families? In a way, they emphasize the inverse problem: for a superordinate category like bird, she wants to find the best or most typical instances. Our problem is the other way around: for a given concrete form like a fork, we want to find the best or most typical invention family. This is a matter of memory organization rather than logic. Even though logic can tell us nothing about the best member of a category (Rosch) or the best family context (here), I believe that our memory organization has better, if not best, members for invention families. Rosch certainly proved this for category membership in an elegant series of studies. I think I can demonstrate its plausibility for family relations as well.

Take the following sentence frame:

< A fork is a(n) _____. >

and substitute into it the list we started with: eating utensil, piece of metal, and so on. For each substitution, rate how well it fits, from 1 = poor fit to 5 = excellent fit. Based on some very informal experiments, I can say that most people produce substantially different ratings for the various substitutions. Some category families provide contexts that are too loose, while others seem just right. Even among the "right" families, different placements produce differences in focus. To say that a fork is an eating utensil (function) may suggest to us different origins than to say it is an instrument with three or four sharp tines (structure). In the first case, we are likely to think of knives and spoons as a context; in the second, of pitch forks and sharp sticks.

It is those "right" invention families that we usually want for an invention's context. We know we have made the right choice when the context is rich in interrelationships and we see a family of inventions that intuitively belong together; and when our placement in a family reveals gaps and suggests extensions and new invented forms. These are the reasons for finding an invention family.

Putting It Together to Find Possible Worlds

We have now looked closely at descriptive categories, as well as concepts like parsing, variation, and evaluation. A good test of understanding is to look at the microworld of some simple object and then apply these concepts.

Our microworld begins with the common safety pin and the fisherman's snap lead, as shown in Figure 6.4. We will assume that the two are related, that the snap lead could have come from the safety-pin idea, even if the actual history may have been otherwise. We begin by parsing the safety pin. Its perceptually and functionally isolable parts

The safety pin and labeled parts.

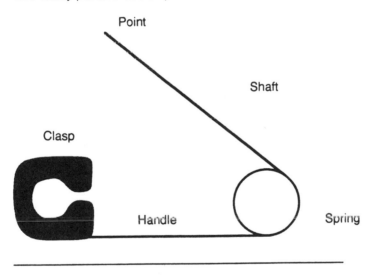

Fisherman's snap swivel.

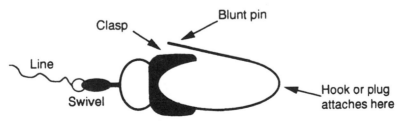

Figure 6.4 A safety pin and a fisherman's snap lead, two siblings in structure but with somewhat different purposes. How can we get from one to the other?

consist of a pin, a clasp, a spring (sometimes a clip on the spring to keep it from separating), and a back (handle) that also holds the parts together in the proper spatial relationship. The safety pin's overall function is to fasten together material, such as pieces of cloth, in a reversible way; the separate pieces of cloth can be alternately fastened or taken apart. The actions involved in using the safety pin include opening and closing it and holding the back to insert or remove it from the fabric it is applied to. Its component materials are a relatively soft metal for the clasp and hard spring steel for the pin, back, and spring. An overall evaluation might say that it is quite effective. But when looking at the components, one might make some negative evaluations. Thus it is sometimes difficult to get the pin and clasp in proper alignment, the pin may rust (if used on diapers, say), and the pin can be dangerous to the one who fastens it or wears it.

To identify the dimensions of variation for the safety pin, we perform a parse and examine the several candidates. The clasp material can be varied. In fact, we can minimize rust if the clasp is made of nonmetallic material such as plastic (many safety pins, especially for diapers, do have a plastic clasp). The large plastic clasp also may be grooved to make it easier to get the pin into the clasp, especially under adverse environments, like in the dark. The potential danger of the pin suggests replacing the pin mode of fastening with another type of reversible fastening, like Velcro or tape. However, this is a dramatic change in the principle of fastening, and it may introduce some subtle tradeoffs like an unsterilizable material (safety pins may be boiled but only special forms of Velcro are heat tolerant).

Other variations come to mind. Does it make sense to dramatically alter the scale of safety pins—and perhaps the scale of the materials they fasten together? For example, with a size increase, might we apply safety pins to building construction and move walls around at will by safety-pinning them to the ceiling? We could create our own designed interior, just as a clothing designer creates new garments initially and temporarily with pins. However, reflection indicates that a scaled-up safety pin is not the best fastener for this purpose, even though the idea is interesting. As another variation, we can try downscaling the safety pin, making it much smaller. When we do this, we find an interesting realized variation, the *swivel snap lead* for quickly changing fishing lures, as seen in Figure 6.4B. It may be viewed as a modified safety pin, one without a sharp point. So the presence of a point must be a dimension of variation as well. This form of the safety pin acts as a quick and convenient interface between the fishing line to which it is firmly tied and the lure that is frequently changed. The example tells us that the kinds of materials fastened may also be varied. All told, an interesting use of the safety-pin idea. Can we scale down the safety pin still farther? To do so requires a different method of fastening it, our fingers being too large and awkward. Since that doesn't seem like a good idea, let's back up and see what kinds of changes in the safety pin are needed to get an effective swivel snap lead.

In the passage from safety pin to snap lead we see a related purpose, holding parts together in a reversible way. But that is a very general statement, and the specific purpose has certainly changed. Instead of fastening different edges of cloth as does a safety pin, the snap lead connects different families of objects, fishing lure and line. Purposes are not immutable, they expand and contract and drift. Here the invention's principle is more constant, fastening two components together in a reversible way by using a pin and clasp. In parsing the safety pin to look at parts, we see that the fisherman's snap swivel has subtle shape changes made to accommodate it to its new role. Then we come to materials. The snap swivel must have noncorrosive parts because it will be in the water. This is probably why some of its parts are made of brass.

Now that we have practiced the mental paths of parsing, variation, and evaluation for existing inventions, how do we move into generating a new idea?

III

THE HEURISTICS OF INVENTION

7

Heuristics
as the Engine
of Variation

Brainstorming's notion that we should generate a lot of ideas and then select from them is a mental Darwinism: Be fruitful and multiply your ideas and then let selective forces reign at a later time. Other investigators, far from the brainstorming tradition, have also favored Darwinian models of the inventive process, at least for early and unsophisticated invention. Their models also emphasize chance variation and then subsequent selection from among those variations. Certainly, chance has a place in invention. However, we should not overestimate its role. Just try to come up with improvements by randomly changing the connections in your stereo or the symbols in your computer program. The role of chance in evolution has millions of years to do its work. Human culture and technology moves much faster; more than chance and accident are involved here.

Another commonly cited source of inventive ideas is pure reason, guided by logic and mathematics. This is the Sherlock Holmes idea. Through an exercise of pure deductive logic the puzzle is solved, the culprit is identified, and the world is in balance. In science and technology, reasoning frequently involves sophisticated logical and mathematical formulations both of which are heavily driven by theory. Together, theory and reason interact to produce new concepts about the ways the world works.

This view is appealing, but it too is a partial answer to how ideas are generated. Anyone who has had contact with a first-rate mind soon realizes that much of the thinking is loose and metaphorical in the early stages, at precisely the time ideas are being generated. The two foremost physicists of the century, Albert Einstein who is responsible for relativity and Max Planck who is responsible for quantum mechanics, substantiate the view that idea generation is different from formal reasoning. Einstein in a widely cited quote maintains:

> The psychical entities which seem to serve as elements in thought are certain signs and more or less clear images which can be "voluntarily" reproduced and combined. . . . [T]his combinatory play seems to be the essential feature in productive thought. . . . [The] elements are . . . of visual and some of muscular type. . . . [These combinations are put together] before there is any connection with logical construction in words or other kinds of signs which can be communicated to others.

And Planck says:

> To be sure, when the pioneer in science sends forth the groping feelers of his thoughts, he must have a vivid intuitive imagination, for new ideas are not generated by deduction, but by an artistically creative imagination.

Both of these individuals were primarily scientists rather than inventors. Perhaps their intuitive starting point is true for science but not for invention? An important inventor Jacob Rabinow, who first brought machine character recognition to the Post Office, has this to say:

> It bothers people when I say that inventing is not done logically. . . . I believe that inventions are an art form and that new things in technology are done in the same way as one would write poetry or compose music. . . .

There is striking agreement among these three important creators about the nonlogical processes in generating ideas. In passages not cited, each also reserves reasoning and mathematical proof for a role that takes place later, after the idea generation stage. Evidently, logical and mathematical reasoning are used to confirm intuitions and ideas sometime after the generation process. Creative thinking is more flexible and much broader than formal reasoning.

How do inventive people really work? Undoubtedly, in many ways. Yet a common one is to generate a few good ideas and rapidly evaluate and develop them. This is not brainstorming because no effort is made to produce a lot of ideas, just enough to play off against one another, and evaluation occurs quickly. The process is not formal reasoning either, because the thinking is loose and the evaluation intuitive. Nor do these inventive individuals spend much early time in chance exploration. Instead of working by blind chance and selection, these creators often *select their next invention steps before they generate them.* Contradictory? Not really, if we focus our idea generation into promising regions of variation. This claim is made plausible by example. Imagine writing a sonnet or a computer program by randomly generating symbols and then selecting the most sensible of the resulting variations. Even if we make the problem simpler by starting with a finished sonnet or a working computer program to revise, we are not likely to improve it in reasonable time by randomly changing symbols. To do this is to court disaster. We must have selection criteria that guide our generative activity into small regions of the space that describe possible variations on our sonnet, computer program, or invention. At the very least, we select a strategy for a given situation before we generate variations, if we are going to be intelligent creators.

So we need another answer to how we best create or generate ideas, a path somewhere between pure chance and pure reason. Call it a *third path*. That third path is through informal heuristics of invention and discovery. As we have seen, a *heuristic* is a strategy or rule of thumb for generating ideas or for solving problems. Everyone is familiar with the concept, if not the name. Thus proverbs and homilies are essentially heuristics of conduct. "The early bird gets the worm." We know this advice doesn't always work. Sometimes the late venturer comes to success; sometimes the early ones fail. However, as a rule of thumb, the admonition to get there first has a lot going for it as a heuristic of conduct. In a similar way, invention heuristics enable us to move much more quickly than blind chance and subsequent selection will allow, and they permit us to operate in vague fuzzy domains where formal reasoning fears to deduce.

The idea of heuristics, informal rules of thumb for solving problems and generating ideas, is an old one. It was first made popular by the mathematician George Polya. Psychologists like Allen Newell and Herbert Simon applied the heuristic idea to human problem solving. And most recently, heuristics have been used to understand the genesis of scientific discovery and recreate the path of important scientific discoveries. But let us see now how the heuristic idea applies to invention.

Heuristic principles constitute the engine that drives inventive variation. When systematically applied, heuristics drive invention faster than any mechanism based on chance variation, at one extreme, and faster than careful reasoning, at the other extreme. A good heuristic lives halfway between chance variation and deductive certainty. An example invention heuristic is to look for an inverse, an undoing action. Thus the marks we make on paper with a pencil often need to be undone in order to edit our prose, our actions. A suitable inverse is an eraser; another is the whiteout liquid we cover typed mistakes with. Similarly, a hammer is assembled with both a striker end for driving nails and a claw end for removing them. In its full generality then, the inverse heuristic says that whenever we have an interesting invention, try next to find an inverse for it, because that inverse may be interesting and worthwhile. Just as the folk wisdom of a proverb is to guide conduct, so an invention heuristic is to guide idea generation.

Here are some other ways of thinking about heuristics. If a description is like a complex noun, with many dimensions, then heuristics are like complex verbs. When a heuristic (verb) is applied to the description (noun or noun phrase), we generate other candidate sentences or phrases. Suppose a situation with several nouns and a verb of combination. One complex noun (actually a noun phrase) describes a simple pencil without an eraser: "a long wooden cylinder encasing lead for the purpose of leaving behind a visible mark." And another complex noun describes a separate eraser: "a piece of rubbery material for the purpose of erasing pencil marks." Suppose our complex verb (or verb phrase) is: "combine those tools or implements that undo the actions of one another." The result of applying the verb (heuristic) to the nounlike descriptions of pencil and eraser is then a combined pencil and eraser—something relatively new, by the way.

If the language analogy of nouns and verbs is not appealing, consider another one: A heuristic is a strategy or procedure for helping us understand a given invention or for changing one invention into another. Heuristics, like most strategies, do not guarantee success; they are rules of thumb. Human intelligence is still needed to decide on the appropriate heuristic to apply and then to determine if the application works.

The Role of Invention Heuristics Applied to a Haystack

Perhaps the best way to understand heuristics is to consider an example, one based on the proverblike task of looking for a needle in a haystack. The search for a needle in a haystack provides a metaphor for search; the target of search, an inventive idea, is the needle; and the search space of possible places to look is the haystack. Searching for a needle is purposive, and to see the role of purpose we will contrast searching the haystack with a simplified account of evolution.

One basic search operation is to take a bit of hay at a time, look for the needle, dispose of the searched hay, and continue until finding the needle or giving up. A *blind chance* approach is to randomly grab handfuls of hay, look for the needle, and throw the hay back. This kind of chance grab and examination of the result is probably the weakest invention heuristic of all. With blind chance, we may very well end up looking at the same hay more than once. In a similar way, evolution makes the same "mistake" on many occasions, as the same lethal gene is mutated. If the haystack is not too big and we have a very long time to wait, this is a way that can produce good ideas—or a successful organism.

A more powerful heuristic is to employ *systematic chance.* Now whenever we look through a handful of hay, we discard it so we will not have to look through it again. Such simple record keeping will speed up our search dramatically, on the average. In fact, we have found a good heuristic: make sure not to search in the same place twice. (I'm assuming that care is made in going through the hay-in-hand so the needle will not be overlooked.) Evolution accomplishes this remembering operation by not letting its big mistakes live long enough to successfully reproduce.

Is it possible to do even better? Yes, we can further limit search by knowing something of our target's properties. That's one of the reasons we spent so much time developing a descriptive framework for inventions. Needles are metal, and metal has a greater density than hay. By getting smart and using good heuristics, we can avoid long searches. Simply shake the haystack to let the needle settle to the bottom. On average, with enough shaking, the needle will fall to the bottom of the stack, and we will find it with a minimum of search. Knowledge, heuristics, and theory make just such reorganizations possible, and they minimize the hay we need to sift through. There is no comparable purposive process in evolution.

Still other properties of the needle and hay will guide us. The realization that needles are made of metal suggests that magnets will be helpful. With a good magnet (like a good heuristic), we can begin to probe the stack. More powerful magnets can search better and bigger haystacks. A bigger and still more sophisticated instrument (heuristic) results in even less search. Hopping to another principle like a big ultrasound scanner or an X-ray scanner provides a complete escape from turning hay. Simply look at the CRT display for the ultrasound trace or at the X-ray plates for the tell-tale signal of the needle. As our metaphorical magnets, sound generators, and scanners get bigger, the sophistication increases and we move from weak heuristics to strong heuristics to firmly based theory.

Knowledge at many levels is at work here: knowledge about the properties of needles, hay, and instruments; heuristics and theory about how to make those instruments; and then skill, theory, and heuristics about how to proceed. But we are not applying this knowledge in a strictly deductive fashion. The reasoning is loose and plausible; in short, it is heuristic reasoning.

That's just the kind of reasoning that fills our everyday actions. Life is moved not by certainty but by the impetus of implicit and explicit heuristics. For example, a motivationally directed heuristic is: "Work hard to get ahead." We all know such an admonition is anything but certain; it is a probablistic thing that says our odds of getting ahead are better if we work than if we do not. But sometimes lazy people do well, and we also know people who are talented and work hard, but to no avail. In short, the work-hard maxim is a reasonable guide to action. Because it can apply to anyone is almost any situation, it is a very general heuristic. Because it does not tell us how to go about our work, when to do it, when not to, it is also a weak heuristic. Without further specification, it will most likely not lead to powerful results. Moreover, other seemingly contradictory maxims can be used to guide our lives—"Be mellow," "Go with the flow"—each of which as its adherents and its contexts of applicability.

This state of affairs is summarized in Figure 7.1 in a power-generality function, a graphic depiction of how heuristic power and generality relate to one another. Heuristic power is roughly synonymous with efficacy in suggesting useful ideas, and generality is related to the number of situations in which you can use the heuristic. In the figure, the work-hard heuristic WH is located somewhere in the lower right-hand corner at point WH which indicates a general procedure that is also weak in its power. A quite different heuristic is the long-division procedure LD that we typically use to divide numbers; it is shown in the upper left-hand corner. The long division procedure differs in several ways from the work-hard heuristic. It is a special form of heuristic called an *algorithm*. For an algorithm, if we follow its steps correctly, we are guaranteed the correct answer to our problem. (For our purposes, an algorithm is a heuristic that works with a probability $= 1.0$ in a finite number of steps.) Therefore, the long-division heu-

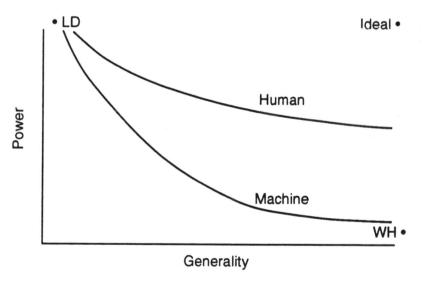

Figure 7.1 The power-generality tradeoff. Human intelligence is much more robust and capable of profiting from ill-defined heuristics than is present-day machine intelligence. WH = the work-hard heuristic, LD = the long division heuristic. Ideal = what we really want in a heuristic, a principle that is both general and powerful.

ristic (or algorithm) is powerful in its ability to generate a good answer to the division problem. Yet its generality is sharply limited. Unlike the work-hard heuristic that can be used almost anywhere, the long-division heuristic can be used only for division problems.

These two examples illustrate the tradeoff between power and generality. Other examples of general and weak heuristics abound: If something doesn't work, shake it or kick it; if the stereo (or TV or CD) is broken, try randomly reconnecting the wires. While these heuristics are very general in the sense they can be applied almost anywhere, they are not likely to produce a successful result. Contrary examples of restricted but powerful heuristics are also plentiful: To add a column of multidigit numbers, start at the right side and use the carry procedure you learned in elementary school; to record on your VCR, execute the steps in your manual in the following order. . . .

Ideally we want our invention heuristics to be both powerful and general, but often such an ideal point in the upper right-hand of the Figure 7.1 is contrary to the nature of the world. To solve problems with computers and machine intelligence, we need specific heuristics (often algorithms) that are powerful, even if they have limited generality. The reason for this is the power-generality curve for present-day machine intelligence (the bottom curve) is very steep in its descent. Even middle-range heuristics, like looking for an inverse to undo something, may not be very helpful for machine intelligence. For example, putting an eraser on the end of a pencil is incorporating an inverse or undoing the capability with our pencil. We all know that an eraser *attached* to a pencil is useful, although the combination is little more than one hundred years old. But unless the notion of specific inverses is already built into a computer program, machine intelligence is not likely to find helpful even such a middle-level heuristic.

The hypothetical power-generality curve for human intelligence (the top curve in Figure 7.1) is much less steep. That means that the heuristic of looking for an inverse may be helpful for human problem solving. Indeed, other heuristics of an even more vague and general nature ("focus on a goal"; "think of a similar problem"; "develop a plan") may be useful for human problem solvers.

In what follows, our approach to invention is based on human intelligence, so we will not worry that the heuristics we use—in the middle to somewhat vague level of generality—are of little use to present-day computers and artificial intelligence. Human intelligence is wonderfully flexible, powerful, and general when it is focused systematically. That focusing is our aim, so let's look at some specific heuristics.

8

Single-Invention Heuristics

Many times an invention's description suggests next steps, by the simple application of heuristics to the description. Let's begin with heuristics applied to a single invention, heuristics that are especially useful in helping us explore and cross the fuzzy boundaries that often separate one invention incarnation from another.

Making Variables

I want to begin with a whimsical story, one without any historical basis. Suppose we are looking at a pack animal. The problem is to construct a transition or path that will enable us to slide gracefully from the concept *backbone* to what we now call a vehicle *chassis*. Such a problem might occur if one were trying, for the first time, to construct a load-bearing vehicle that is modeled after a pack animal. What should we do?

A good beginning is to examine the animal model for parts that may help in the construction of an invented counterpart. Useful parts include the pack animal's backbone and the ribs sticking out at approximately right angles; this is what we balance the load or pack on. The next step is to construct a description for backbones and ribs, and to create variables or dimensions corresponding to these fixed animal characteristics. The description for Backbone might include as dimension and value pairs the following: material: bony; structure: relatively rigid; axis: elongated; placement: interior dorsal region, that is, the top of the animal; and function: supports load. The transition to the invented counterpart, which we will call a *chassis,* then becomes: material: *wood;* structure: rigid; axis: elongated; placement: *exterior lower body,* that is, the spine is now down low; and function: supports load. The italicized values differ from the original values for backbones. In this simplified example, the principal differences between the backbone and chassis are the values for the material and placement dimensions. To get from Backbone to Chassis then becomes a matter of treating some aspects of the backbone model as dimensions or variables that can take on a range of possible values: "bone" or "wood," "interior dorsal region" or "exterior lower body." By making these substitutions, we have slid from a pack-animal model into the idea of a chassis. Our first result is in Figure 8.1A.

Creating a variable in place of a fixed or constant feature is a common mode of invention and idea generation. In the present case, the location of the backbone floats—that is, it takes on the different position values "dorsal" or "ventral." Notice that certain potential dimensions (and corresponding values) such as odor (bad) or color (white) are not of major concern in the path from backbones to chassis. For vehicle design, we drop out the odor and color aspects by treating them as irrelevant for carrying loads. The important idea here is elon-

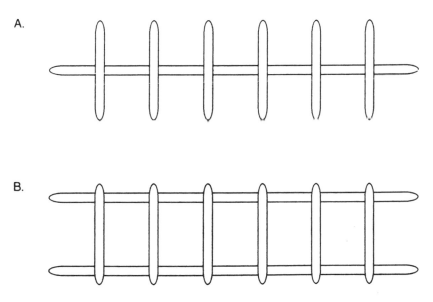

Figure 8.1 Backbone to chassis, a whimsical progression. (A) First attempt at modeling a chassis on backbone and ribs. (B) A refinement, duplicating the spine, to correct the stability problem. The result is the sledge, perhaps the first vehicle.

gated structural support around which ribs may hang or be attached. All of this suggests a principle:

> *The Make-Variables Heuristic* Make or identify potential dimensions of variation, that is, create dimensions or variables where previously only fixed features or constants existed. Then begin to change the values of those variables to more closely approximate the desired goal, as determined by evaluative criteria (like the ability to carry a heavy load).

The mathematician-philosopher A. N. Whitehead has argued that the idea of a variable applied to mathematics is one of history's great intellectual achievements because it easily represents abstraction and generalization. In the present context, this simple heuristic, like the others to be offered, supposes an inventor with some knowledge and sophistication. I doubt very much that the rote application of such a heuristic produces much of interest. In the power-generality tradeoff, the make-variables heuristic is quite general and low in power. Yet in the right hands, it can be very useful. The idea of looking at a situation or an invention for what can be changed takes time to develop, and knowing when to apply the heuristic takes subject-matter knowledge. Once more, heuristics of this level will not do much for contemporary machine intelligence, but they are often a good starting point for human intelligence.

Repeated Elements

Let us take up once more the construction of a vehicle. So far, by using a pack-animal model, we have designed a vehicle with a single supporting spine and some wooden ribs running off to the sides, as shown in Figure 8.1A. It is not satisfactory. Whenever we put a pack on top, the pack slips to the side when the vehicle is pushed or pulled, and the ribs on that side drag on the ground. The overall evaluation is poor for a single spined vehicle. Notice that a similar problem occurred earlier for a possible ancestor of the fork, the pointed stick: A large piece of meat speared off the exact center of gravity with the pointed stick was unstable when turned over a fire. That problem was solved by increasing the number of pointed ends on the stick, and in one of our stories it led to the multiple tines on a fork. A similar procedure solves the vehicle stability problem, and a second spine is provided, as in Figure 8.1B. The resulting vehicle is very close to a sledge, certainly one of the first human vehicles that was used on dry land, and closely related to the sled used on snow and ice. That the stability problem of the fork and the single-spined vehicle responded to the same solution, repeating one of the components, suggests an important special case of the previous heuristic:

The Repeated-Element Heuristic Once an interesting component is discovered (the spine or backbone), try copying or repeating it as often as necessary. It may be a suitable building block for more complex inventions.

When taken out of context, these heuristics of making variables and repeating elements seem so general as to be meaningless. Still they come into their own when combined with one another. An analogy helps to make this point. Looking for a needle in a haystack is facilitated by shaking it (a weak heuristic); it is also facilitated by using a small magnet (also a weak heuristic procedure). But when we combine the two, shaking the stack and using the magnet at the same time, the power of these weak heuristics increases rapidly.

By creating variables and repeating elements, we greatly increase the variations that can be played on a given invention theme. As we change the values of variables, we slide across boundaries, moving from one invention to another, as from backbone and ribs to vehicle chassis. In the course of indiscriminate variations another problem is uncovered as well. How can we choose from among all the possibilities open to us in sliding about and around an invention? In a way, we have already answered this problem in two ways: Our inventive activities profit from having a purpose to guide us and a set of evaluative standards to keep us on course. Without both purpose and evaluative standards, we are inventing by Darwin's rules: random variation and survival of the fittest. As we will see later, much of invention is conducted in such a way as to steer around these slow and stark Darwinian realities.

Fine-Tuning

I have an ordinary claw hammer in my hand. The handle is about twelve inches long and the claw arc is turned back toward me as I grip the handle. Why these configurations and dimensions? Surely the handle could be longer or shorter, and the claw could have a different angle.

The previous heuristics—making variables and repeating elements—revealed the range of design options. But once we realize that numerous options are usually available, we must select from them. Tweaking or adjusting a variable to find a better value is one of the most common forms of invention. We will call that activity *fine-tuning*. It may be routine, or it may involve heroic search to find just the right combination.

We can illustrate the idea of fine-tuning by returning to the hammer. Some possible hammer variations are presented in Figure 8.2. Why is it that some seem preferable to others? Perhaps it is because of convention; it has "always been done that way." Or perhaps it is to

Hammer Handle Lengths

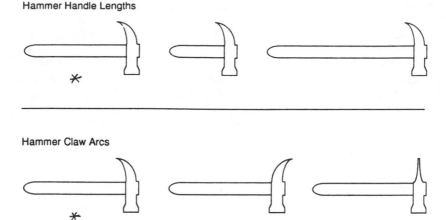

Hammer Claw Arcs

Figure 8.2   In search of the best combination of hammer features. The hammers marked with a * are conventional hammers. Strange happenings occur when the length of the handle or the angle of the claw is changed. Again, the conventional hammer is an example of intelligent design.

satisfy some evaluative standard, like minimizing effort or maximizing impact. Or perhaps to afford a compromise between competing evaluative criteria that involve tradeoffs among one another. Let's explore the last answer.

Tradeoffs are made in any design because of competing requirements. For instance, the length of the hammer handle must balance the competing interests of force and accuracy: Other things equal, a longer handle gives rise to greater velocity at the hammer head and a greater corresponding impact. However, longer handles are also associated with lesser accuracy in striking a target. Hence the actual length of the hammer handle involves tradeoffs of at least these two evaluative criteria: speed and force attained through a long handle, and accuracy attained through a short controllable length of handle. In addition, if the primary purpose is to pound nails, then one hand must hold the nail in position, while the other hand and arm wield the hammer. Because the nail-holding arm is just so long, the length of the hammer-wielding arm and the handle cannot exceed the holding length, so an additional constraint is involved. Recognizing these performance requirements and constraints, and how they interact, is a large part of successful design and invention.

Similar reasoning underlies the arrangement and arc of the hammer's claw. It could be at the handle's opposite end, as a completely separate feature. But then it gets in the way of the arm, and it changes the balance of the hammer—probably adversely. It also requires more material and expense to manufacture. Even the normal placement of the claw and striker in the same head presents subtleties when we consider the claw's arc. The arguments for the conventional arrangement of the claw in Figure 8.2 include the advantage of pulling toward us with our body weight helping as we lean back. If the claw arc angle is turned away from the handle (as in the figure), the leverage is poorer; and it is more difficult to see the nail head when we push the claw-end of the hammer away from us. Also the claw protrudes beyond the hammer so it scratches surfaces when we pound. Considerations like these suggest:

> *The Fine-Tuning Heuristic*   After the right parts are in place, try to rearrange or tweak them for better performance. It is usually possible to improve something by getting all the variables in just the right configuration. To do this sensibly, it is important to have definite evaluative standards in mind.

In some cases, there may be an optimal configuration of variables, the one best solution. Often in mathematics there is one best value of a variable. In invention and simple technology we are not usually looking for a best or optimal arrangement, just something better than we now have. It is almost always possible to improve something. We can keep altering the variables until we don't get any more improvement or until its too expensive to tinker with any more changes. Such fine-tuning presupposes evaluative criteria on which people agree.

Also, notice that to strive for continual improvement is a general and vague heuristic as well as an obvious idea. But perhaps not, because its importance and systematic use have been apparent only to minds of the last few hundred years. That an artifact can always be improved is both a false idea and a radically useful one.

Variance Control

The average annual temperature in Hawaii is about the same as that in high plains areas of the United States. Yet the temperature in Hawaii is fairly constant, in the 70's, while that of the plains varies from sub-zero freezing in the winter to over 110 degrees in the summer. The average temperature in either location is fine but the range of temperature can kill. To survive, we must control that range.

When we talked about fine-tuning, we were concerned with finding better values of a variable. Now we generalize on the idea of the *value* of a variable and come to the notion of controlling a variable's *range of variation*. In a related idea, we may want to take something that is concentrated in time or space—like the energy from a lightning bolt—and *spread it out evenly*. The importance of these ideas is pervasive; and range or *variance control*, as we shall refer to it, comes in many contexts, from the simple to the complex.

One of my favorite examples is a very old invention, the shaduf, which is shown in Figure 8.3. The shaduf is a mechanical device used in the Middle East to lift water. It has something in common with a stairway, a button, a zipper, a thermostat, an irrigation system, a memory buffer, and a nuclear reactor. That common characteristic, termed *variance control*, will be revealed as we deal with each of these devices or systems.

The shaduf is a long stick weighted at one end; the other end is attached to a bucket that dips into water at a lower level, as shown in Figure 8.3. The stick is "hinged" by a forked stick in the middle that serves as the fulcrum for the lever stick. The question we address is why have the weighted end? The weight makes *lowering* the empty bucket into the water more difficult because we must raise the weight to get the empty bucket down in the water. Still, when it comes to *lifting* the water-filled bucket from the river, the weight helps us: We need to apply less effort on the lift-end than if the weight were absent. A tradeoff on the variable of effort is suggested: Spend more effort in getting the empty bucket into the water and gain advantage in lifting the full bucket. This tradeoff makes more equal the effort in the cycle of action: lower, fill, lift, swivel, and empty. We gain the ability to lift bigger buckets out of the water by being willing to exert greater effort in lowering an empty bucket with a counterweight at the other end.

Said in still another way, the counterweight makes for less variability in effort, with the exertion required for lowering the empty bucket into the water approximately equal to that for lifting a full bucket from it. We have spread our efforts evenly in time instead of having to alternate between easy and hard cycles. This principle of equalizing a quantity, such as effort, is an example of variance control.

Variance control also explains why we get to the top of tall buildings not with a single bound but through climbing a winding staircase.

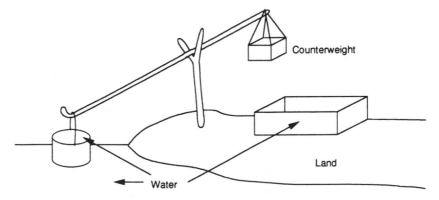

Figure 8.3   The shaduf, an ancient water-lifting device for equalizing effort on the down stroke and the upstroke—an early example of variance control.

The staircase allows us to replace one huge effort—that we may not even be capable of—with a series of small efforts spread out over a longer time. Said in another way, the staircase allows for incremental progress toward our goal of getting to the top, and it also provides convenient points of stability where we can rest and not have to worry about rolling backward. Variance control, in the sense of equalizing effort over a span of time, is implicated again.

Variance control operates not only in inventions involving effort or work but on other dimensions as well—and here is the link between the shaduf and those other devices, like clothing, that we promised to bring into relation with one another. Why do we need to wear clothes and animals do not? And what are the effects of wearing or not wearing them? The *why* we ask relates to the functional reason. The functional gain of clothes over fur and feathers is the ability to rapidly adjust to new temperature conditions—and to keep the range of temperature next to our bodies within tolerable limits. An advantage of this is that clothing can be put on or taken off, while fur cannot. This means a greater spectrum of climates to which the human is thermally adaptable. Although particular animals may vary the thickness of their summer and winter coats by shedding and the like, the change is much less rapid than it is for putting on and taking off clothing.

Variance control, then, also operates in adapting to changing temperatures. We put on and take off clothing to keep the body surface at about the same temperature. We track our thermal needs by using clothing as an interface between our bodies and the external climatic conditions we find ourselves in.

But the relation of variance control to clothing is still more intricate. Clothing fits on the body in a variety of ways. Two of the more common include the tunic, essentially a tube with arm and head holes, and the open-in-front garment, like a jacket. Which of these two forms allows for greater control of temperature variance? The tunic is close to being an ON or OFF garment (although it is possible to roll it up a bit). So its range of temperature modulation is quite limited. Often we control temperature when wearing tunic-like garments by having several layers, each of which can be pulled off like the successive layers of an onion. The outer core might be a parka that slips over the head, the next layer a sweater that slips over the head, then a rough shirt, and finally an undershirt—all in tunic form, and each one separately removable. Multiple tunic garments work well to control temperature, so long as we don't have to carry the successive layers as we take them off or put them on.

The open-in-front garment works better for controlling temperature range at the body surface. But it presupposes a reversible fastener—a tie or set of ties, hooks-and-eyes, buttons, zippers, or Velcro.

Each of these fastener systems can be thought of as a way to provide a reversible and adjustable thermal interface between the climatic conditions and the thermal needs of the body. In short, each fastener system is a way of controlling temperature variance at the surface of the body.

So far we have considered variance control in the context of work, effort, and temperature. Is the concept still more general? Let's return to the example of the shaduf and look at it from another perspective. A scarce resource like water must be husbanded carefully. To husband something means to store it during times of plenty and release it selectively during times of want. Here we want to spread a quantity such as water over a longer interval and in selected places. Once again, this is variance control, but now over time. Seen in this way, the shaduf is an element in a system of variance controls, one that we call irrigation. The system requires capturing water, storing it in reservoirs, then releasing it through gates or valves, sometimes lifting it (where the shaduf comes in), and moving it through channels. Certainly, the focus of irrigation is controlling and distributing the uncertain presence of water. Any undesirable, variable stage in this system is fair game for human invention, not just the transitions in lifting water from one height to another, but also the storage, the inevitable losses that occur between input and output, and the recording or metering of the resource as it passes from one place in the system to another. The principle of variance control is present at many of these points. Indeed, we build dams and tanks to form reservoirs that will allow for the controlled release of water. Related inventions such as the pipeline allow us to spread water to selected areas, and water meters provide an accounting of the variable use of water.

Now for a more modern example. Much has appeared in the news about controlled nuclear reactors. What does that have to do with the shaduf and variance? In a way it is the same problem. People have known for some time how to produce *uncontrolled* fission; it is called a nuclear bomb. Yet to use that energy for more salutary purposes, we must control it by spreading it out over time. We want to replace a brief spike of intense uncontrolled energy with a much lower level of energy spread over a much longer interval. That is essentially what nuclear reactors do.

Just as energy is an intangible that is often in need of control, we can think of another intangible to add to our list: information. Like a violent rain that will run off unless stored in a reservoir, information comes to the senses more rapidly than we can process it. So evolution provides our senses with storage buffers, stages of information processing, and more buffers—a sequence in which the fleeting nature of information is prolonged so different mental operations have time to be performed along the way. This is doing to information what was done with water, irrigation, and energy—buffering it and regulating the flow for maximum effect. In short, making more even the flow of information by controlling its variable rate past a given stage.

Not just biological systems control the flow of information. The buffering, ordering, and selective retrieval of information, done by sophisticated computer systems, provides an example of less exploited quantities and devices. There is simply too much information to deal with. When this happens, we can control the flow by shutting down completely or installing a selective filter. An example of filtering information is provided by the electronic newspaper, now available at places like MIT's Media Lab. Type in the descriptors that interest you, terms like *variance, control, quantity, information.* You are then presented with a variety of current news items that have embedded in them your descriptor terms. Ideally, as you select some items for more extended reading and reject others, the system learns your preferences, and it tailors the retrieved items to the length and detail you desire. Of course, if a term appears that you want to know more about, simply click on it, and you will be off on a side trip. You can do this as often as you wish. The usual name for such a system is *hypertext*—a network of electronic linkages that can take you to related items of interest. If it is a sophisticated hypertext system, it will begin to learn and track your preferences as it guides you.

A problem with hypertext is that it can only seek out articles containing the terms we give it. Thus if we had used as descriptors *variance, control, quantity,* we would not have gained access to articles on the shaduf. Why? Because people do not think of the shaduf in those terms. For analysis based on underlying principles, hypertext and other electronic retrieval systems are a long way from replacing the organizational power of human intelligence and memory. We can only speculate on what intelligence an electronic retrieval system needs to search text for deep principles not explicitly mentioned.

We have reached a surprising result. At a very general and deep level, controlled information flow in hypertext, energy emanating from a nuclear reactor, and water flow as maintained by a system that includes the shaduf, an invention used on the Nile thousands of years ago, are all answers to the same problem: variance control.

All of our examples suggest:

The Variance Control Heuristic Seek out variances that need to be controlled, from concrete quantities like water to abstract quantities like information. Then try to find improved methods of control in order to minimize range or to spread it out more evenly.

Feature Addition and Deletion

Last week I saw the makings of a terrible accident. A runner with earphones and a Walkman passed me. He ran into the intersection at full stride. Brakes screeched and a car swerved, just narrowly missing the runner who was now doing an elaborate contorted dance to overcome gravity and motion. The driver went on by shaking his fist, and the jogger got to the other side of the road where he sat down shaking. I ran over, "Are you okay?" He looked blankly at me, took off his earphones, and asked what I had said. We talked for a few minutes about his close call.

As much as any invention, Sony's Walkman radio allows us to build a personal bubble of our own reality around us, although sometimes other realities like swerving cars intrude. The Walkman also provides a case history of inventing by adding and deleting features. And it illustrates a social invention that contributes to the development of physical invention, a person playing the role of a Walkman/Integrator between laboratories.

Before taking up our story, some background is in order. The transistor radio was one of Sony's early marketing successes; it was highly portable and could be taken anywhere. Also portable was the Sony Pressman, a small mono tape recorder of high quality. Record and playback functions were present, and playback was monitored through a single speaker. The targeted market included reporters, interviewers, and everyone who needed a tape recorder that was portable and had reasonable sound quality.

A research group at Sony tried to make the Pressman into a stereo recorder by adding stereo circuitry. But the bulky circuitry then in use allowed enough space for only playback and not for record. The group met resistance. Whoever heard of a tape recorder that did not record? Also, the small speakers needed for portability didn't produce quality musical sound, a poor evaluative outcome. And if a second speaker were added, the small configuration didn't allow for stereo separation. To fix the problem of stereo separation, one or both speakers could be made detachable, but that adds bulk and decreases portability.

Honorary Chairman Masaru Ibuka, the retired head of Sony, entered at this point. Management had not known what to do with a retired chairman who persisted in showing up for work, so he wandered from one lab to another. The wandering made him familiar with the various research projects and problems. One lab he visited was trying to develop lightweight high-quality headphones. Unlike the heavy earmuff headphones that wrapped around the whole ear, these headphones had to be light enough and comfortable enough to wear for a long time. The new headphones also produced excellent stereo sound. No one thought of them as portable—yet.

Later, when Ibuka visited the lab that was working on the stereo version of the Pressman, he saw both its potential and its problems. He suggested dropping the speakers and adding the lightweight headphones. Not having to power a speaker, the batteries would have a longer life. To make the device even smaller and lighter, he suggested forgetting about the record function altogether.

Skepticism greeted his suggestions. His approach contravened what Sony engineers felt was the normal direction of product development, the incremental addition of functions. (Actually, this is not the only direction for development. Specialization and simplification of products are also time-honored development patterns.) Other objections followed: What if several people wanted to listen at once? What if you wanted to record something? Still, Ibuka's idea of dropping the speakers and the record function offered several advantages:

- A decrease in cost, because the speakers and the record function were eliminated.

- A decrease in weight and size, also because of fewer functions.

- An increase in portability, because of the reduced weight and size.

- An increase in sound quality, because small speakers cannot compete with good headphones.

- An increase in stereo separation, because headphone stereo is superior to speakers attached to a small tape player.

- An increase in battery life, because headphones use less power than speakers.

- An increase in privacy, because headphones shut out external sounds better than speakers do—a personal bubble is created around oneself even in the most crowded situations.

The critical inventive ideas consisted of feature additions and deletions to the Pressman recorder; the addition and deletion of features are common routes to invention. The highly specialized headphones (very light with high fidelity) were added to the tape player. This is equivalent to adding parts and dimensions to an existing invention. The speakers and the record function were dropped. This is equivalent to dropping parts and dimensions. These feature additions and deletions contributed to the Walkman's light weight, small size, and portability. The result was a good fit between human and machine.

Some forms of the Walkman, and its equivalent produced by rival manufacturers, now have tape *recorder* functions and a radio as well. These added capabilities have resulted from ever-increasing miniaturization and integration of electronic components. Most of the electronics for a radio can now be packed on a single integrated circuit chip.

The case history of the Walkman, then, suggests two very general invention heuristics:

The Feature Deletion Heuristic   See what features or dimensions you can delete from an invention and still have a viable product. (*Example:* The record function and the speakers.)

The Feature Addition Heuristic   See what other features or dimensions you can add to a given product. (*Example:* The ultralight and comfortable headphones.)

As the Walkman illustrates, these heuristics can work in tandem. However, there is more here than just randomly adding or subtracting features. One must have a hunch about what features could likely be dropped without much loss and which added with much gain. For the Walkman, evaluative criteria such as portability and stereo fidelity triumphed over having record and speaker functions, and ultimately guided the removal and addition of features.

The development of the Walkman also made use of a potentially important social invention, an important addition to the process of product development. Evidently, the different labs at Sony did not regularly communicate with one another. When Honorary Chairman Ibuka wandered from lab to lab, he acted as an unofficial integrator of different technologies. He was a "research and development walkman," or perhaps we should refer to his role as an *Integrator.* Rarely do complex organizations have someone in a role like this. At my university we have hermetically sealed units. Few ideas or threads pass from one department to another. What if we had several Ibukas, distinguished retired people who wander from one lab to another with the charge of connecting the threads of ideas across administrative boundaries? Why not use the wisdom and communicative talents that have been delicately honed over a lifetime? The Walkman-Integrator is no longer doing active research and should not threaten the ideas of others. He or she can provide needed threads or connective tissue between encapsulated bureaucracies and administrative units. The idea of the Integrator is so simple and radical that it might work exceedingly well. Or might it frighten forces that feed on encapsulation?

So *Walkman* not only names a product, but it metaphorically describes a role: the *Walkman/Integrator,* one who walks across the boundaries of departments, with the idea of weaving together the threads of separate imaginations.

9

Multiple-Invention Heuristics: Linking

When I was younger, I always had a stationwagon. Not because I liked the ride or the way it responded or the exposure of cargo, but for reasons of economy. Ideally, I needed a sedan for the comfort it affords on longer trips and a pickup truck for all the junk I wanted to haul around. The stationwagon was a compromise, a kind of functional blend or interpolation between the sedan and the pickup truck. That's an interesting new idea, that one invention can be between two others.

So far, we have considered heuristics to focus on a single invention. Most of these heuristics can also be applied to more than one invention at a time. However, let's now concentrate on heuristics that come into play primarily when two or more inventions are involved. These heuristics use the ideas of interpolation and extrapolation, making connections or links between inventions, and then connecting them to a missing purpose.

Interpolation and Extrapolation

Previously we noted that two fundamental problems confront the inventive mind: Not having enough ideas, so not knowing what to do next; and having too many ideas, so not knowing what to do next. That next step can often be taken by applying interpolation and extrapolation heuristics, two forms of search. Both require as a starting point at least two different states of an invention. Call them *early* and *late,* but dimensions other than time will also work; for example, *simple* and *complex* serve equally well.

For interpolation, we seek as a next step or target an invention partway between the early and late forms. For extrapolation, we seek as a next step an invention that continues the trend from early to later forms of an invention. Because of the often immense number of next steps that can be taken in the space of possible inventions, the interpolation and extrapolation heuristics help guide us in our journey through that space. They are based on the presumption, not always true, that we are safe in being guided by a trend.

First, let's consider interpolation or, if you prefer, blending. A recent success story in automotive design is the Chrysler minivan. It is intermediate in size, somewhere between a regular sedan and a full-sized van. How might the idea for it have developed? One explanation is through an interpolative process in which we take as input the descriptions for the sedan and for the full-sized van. As output, we determine a product that will be part way between the two. Such a product is something like the minivan; it has much greater capacity than a sedan but is not so large or expensive as a full-sized van. A similar automotive example is the El Camino, a vehicle that is halfway between a car and a pickup truck.

A different example introduces the ultimate in interpolation, the integrated adjustable invention: the crescent wrench, the zoom lens of the camera, the recliner. The alternative to these integrated forms is to have several different sized wrenches, or several lenses with different focal lengths, or a separate couch and chair. However, with an integrated form, the strategy is to have a *single invention* that is adjustable. By designing a crescent wrench or a zoom lens or a recliner, we produce tools and furniture with indefinitely many interpolative states, whatever the occasion demands. All of these interpolative states are readily available for use by an act of simple adjustment. The wasteful processes associated with having many wrenches of different sizes (looking for the right size, putting down the present one, and picking up another one) are largely eliminated. And portability is enhanced at the same time. But as common with many combination inventions, there is also a downside: The crescent wrench makes contact with only two edges of a nut, so it is much easier to strip the head than with a specialized wrench that encloses the nut and applies forces to each of its faces.

Another example of interpolation comes from utensils. We can conceptualize the evolution of the eating fork in different ways. An interpolative account has one of the original invention states as a large pitchfork-like implement; and the other invention state as a very small implement like an olive fork. By interpolation, we can arrive at an intermediate form, something like a regular eating fork. The account is oversimplified, because not only the scale changes, but the function as well. Pitchforks are not used for eating, and olive forks are usually used for serving rather than direct hand to mouth eating. So by this account, in addition to scale changes, a new function such as eating is also added. Yet adding new functions is a common form of invention.

A more difficult interpolation or blending is to combine forms of a different shape. A ubiquitous eating implement of fast-food restaurants qualifies. The *spork,* as its blended name implies, is a structural interpolation of a spoon and fork. Of course, one may say that it doesn't work very well, but the intent behind it makes it a good example of interpolation.

Now let's consider a different heuristic, *extrapolation.* Here we start with the pitchfork as the "early" state and the ordinary eating fork as a "later" state. Then we ask what is produced if we continue the trend from larger to smaller. In time, we may develop something like an olive fork, a device that continues the size trend from pitchfork to eating fork to something new and even smaller. If history ran the other way, from olive fork to eating fork, we can extrapolate to a larger version, with the outcome a pitchfork. Again, we have oversimplified, because once we produce a new forklike implement by extrapolating to either end of the size scale, we will need to find a use for it.

A different example of extrapolation allows us to escape from the problem of finding new uses for the results. Extrapolation along evaluative dimensions is a very common invention process. Most inventions can be improved. The idea of looking at the world with the idea of improving it is one of the well-springs of invention. It makes its implicit appearance in the long tortuous development of stone tools (producing a longer and a sharper cutting edge with a better human interface), and it makes an explicit appearance in sophisticated pharmaceutical manipulations in which we try to move from a drug that is short acting and requires many administrations to a version that is long acting and with a very smooth release of its effective agents over time, perhaps through a time-release capsule or a skin patch.

All of this suggests further principles to guide our inventive activity:

The Interpolation Heuristic Look for intermediate states in a line of invention development. Some of those states also may be useful. At the very least, we may see a basis for classifying an invention family.

The Extrapolation Heuristic Look for trends in a line of invention development. Try to continue those trends to a next step. To do so may require the finding of new functions or purposes. Or it may simply be the idea of continuous improvement along a dimension that guides us.

Linking and Finding New Purpose

The Allied troops in the Persian Gulf War did not deal exclusively with high technology, as the newscasts would have us believe. Very inventive *low technology* was also practiced. In one case, the troops used pantyhose to keep sand out of the air intakes of tanks. Evidently, the standard mesh covering was not tight enough for the very small grains of sand. One can grasp the reasoning here. Look for something that is flexible, with many narrow openings that are wide enough for air but small enough to keep out sand. In another related case, the troops used condoms over the end of gun barrels to keep sand out. The reasoning behind this ingenious application is left to you.

In both cases, the need or application probably came first, and then there was a search for an already existing invention or material. This is a common form of materials problem. Now computer database programs can help solve materials problems like this. Such a program has data fields like *strength, weight, rigidity,* and the like. A person who is working on an invention will find a need for a material with certain features, perhaps *strong, light,* and *flexible.* When these features are entered into the database, a search is begun and all materials that come close to having the specified feature set are printed out. For nonroutine problems, however, like those faced by the troops confronting fine grains of sand, no computer programs exist, and it is necessary to think through the problem.

A different form of materials problem is to *start with a material and find a new purpose for it;* the limiting case is to find a *first purpose.* I think this is a relatively uncommon form of invention, working from the opposite direction of that seen in the Gulf War, but it has a good deal going for it.

Some people think this kind of invention is infrequent. They may argue that we usually have a purpose in mind when we invent rather than starting with an invention and then looking for a purpose. Yet we frequently do find new purposes, and I suspect that most inventions end up with numerous uses different from those originally intended. Often our purposes change to mesh with our tools and instruments, and their capabilities. Certainly new scientific instruments lead to new purposes.

However infrequently we invent by finding a new purpose, it clearly is a mode of invention needing more emphasis. To take only one example, manufacturers must find new uses for their products. To do so is to increase market and add to the life cycle of a product by doing little else than advertising those new purposes. In addition to its cooking uses, baking soda can be used for deodorizing refrigerators and cleaning teeth. If we can find new purposes for existing inventions, we will serve up the freest lunch that technology has to offer.

To explore this question of finding a purpose, let's look at the search for applications associated with the Post-it adhesive. If people are asked for the properties of a perfect adhesive, likely they will say that it should set hard and fast and strong. Enough of this adhesive should hold the world together. But suppose a new adhesive is discovered that is just tacky and also reusable or reversible? Those are precisely the properties of the Post-it Note adhesive.

Let's trace the idea. In 1964 Spencer Silver, a chemist for 3M, was working on a project designed to find polymer structures suitable for adhesives. (Polymers are long chemical molecules from which things like plastic are made.) According to Silver, rather than rely exclusively on existing theory, he put together some unusual combinations and amounts of simpler precursor molecules:

> People like myself . . . get excited about looking for new properties in materials. I find that very satisfying, to perturb the structure slightly and just see what happens. I have a hard time talking people into doing that—people who are more highly trained. It's been my experience that people are reluctant just to try, to experiment—just to see what will happen!

During his molecular tinkering, Silver found a substance that was weakly rather than strongly adhesive. It would only "tack" two surfaces together. In addition, it could be sprayed on surfaces. As Silver said, the

> material was more "cohesive" than it was "adhesive." It clung to its own molecules better than it clung to any other molecules. . . . So if you sprayed it on a surface . . . and then slapped a piece of paper on the sprayed surface, you could remove all or none of the adhesive when you lifted the paper. It might "prefer" one surface to another, but not stick well to either. Someone would have to invent a new coating for paper if 3M were to use this as an adhesive for pieces of paper.

The promiscuity of the Post-it adhesive would have to be regulated before anything else could be done.

Popular belief has it that Silver was just extraordinarily lucky in finding his adhesive. According to this view, his discovery was accidental; and many, if not most, significant inventions come from chance variations or accident. Such a view is unlikely because the space of possible chemical compositions to search through for useful adhesives is exceedingly large, and Silver's finding can hardly be labeled accidental in the usual sense. Within limits, he knew what he was looking for (adhesives), and he had a defined, although large search space, in which he was looking—some promising chemical combinations. Although existing chemical theory might not have sanctioned his empirical method of search, his method is wrong only if the existing theory will take you directly to your destination, or else tell you that there is no way of getting there. Certainly, in the 1960s neither of these alternatives was the case.

There was another obstacle to developing applications for the Post-it adhesive. Historically, the driving evaluative standards for adhesives have been toward ever stronger bonding in ever briefer periods of time. Silver's adhesive was only weakly bonding, however, and that is what makes it interesting: it is reversible, it can be undone (it has an inverse), and in addition it is reusable. To separate two attached surfaces, just pull them apart. What the adhesive gives up in strength of bonding, it gains in reversibility and freedom of placement. But once a reversible adhesive has been found, one that is only tacky, what shall it be used for, what will be its purpose? This was the question that perplexed Silver and his colleagues at 3M. It forms our basic problem of finding a good application or purpose for a material.

One of 3M's first attempted applications was a sticky bulletin board. It was treated with a substrata to prevent promiscuity, so the Post-it adhesive would adhere to the bulletin board rather than coming off on anything stuck to it. The board was covered with the adhesive, and people then put paper materials on it and took them off easily. The product was something less than a commercial success. In addition to notes posted on the board, one can imagine a collection of dust, dirt, and dead flies. The breakthrough came when another 3M chemist, Arthur Fry, was singing in a church choir. His hymnal was filled with paper bookmarks that were constantly falling out. The idea came to him that if the bookmarks were treated with some of Silver's adhesive, they would stay securely in place. Fry returned to 3M, mixed up the adhesive, and applied it to paper bookmarks. He realized that the best use of the adhesive was to attach paper to paper. The Post-it Note was born.

Again, the popular belief was that this application came to Fry in a flash of insight, and that much if not most of invention results from these lightning bolts of insight. A contrasting explanation is based on Fry's having heard Silver's presentation on the adhesive. Fry then thought about applications of its properties. So even if we accept the flash-of-insight explanation, it was only a small part of his thinking and an even smaller part of the overall development of the Post-it Note. A chemical substratum for the Post-it Notes had to be developed to control the adhesive's promiscuous tendencies; a major effort followed to develop the machinery and processes to manufacture the Post-it Note. Finally, the product required novel marketing strategies. At least the marketing strategies were novel for 3M. The Post-it Note pads had to be given away at first.

For our interest, the big problem for Post-it adhesive was finding a purpose for it—in this case, finding a first purpose. Let's see if we can find a systematic way of doing this. If we can do this in a general way, then we will surely enjoy some free lunches from already existing inventions and technologies.

The procedure that we follow is reminiscent of that in Plato's dialogue *Meno,* in which a youth ignorant of geometry is lead to geometric truth by a series of questions from the skilled questioner Socrates. A suitable purpose-discovery heuristic will serve the role of Socrates in interrogating our own memories and in helping us retrieve the knowledge we need in arriving at an application for a new material. This will turn out to be a long and complex heuristic, closely related to the scheme we have already used for describing inventions.

> *The New Purpose Heuristic*   Begin with a listing of the known properties of the new material. This is equivalent to constructing a description for the material, and then using the descriptive categories we developed earlier to fill in the particular properties of the material. In the description, the function or purpose category will be empty; this is what we seek an entry or value for. To find candidate entries, look for applications that require one or more of the properties that match those of our material. To do this, one searches through memory for a list of known inventions that currently use a material with related properties. In that list, we note inventions that have an overall useful function combined with a poor evaluation for their material's category. Next, the suitability of the new material is tested against the materials used for the existing inventions and checked for a better evaluation. When we find a good match between the properties of our new material and the existing invention, we just kidnap the purpose of the existing invention. Our new material has been shown to be as good or better than that of the existing invention.

If that didn't make sense, don't despair because an example is on the way. But first note that this is a heuristic, so by definition it is not guaranteed to work. One reason that it may not work is that the necessary information is missing from memory. This will be true if it is a problem of knowledge, a matter of simply not knowing the required information. Still it makes sense to first search memory before looking in the world for the right information; saving search effort and time is what the New-Purpose Heuristic is all about. We will now try it on the Post-it Note adhesive to see if the heuristic will recover the adhesive's *known use.*

First, we ask what are the properties of the material? Table 9.1 lists some of the properties of the Post-it Note adhesive. For now, consider only the bonding properties of the adhesive.

Next we ask for the dimensions of variation behind these properties. Presumably, they are strength of bonding, reversibility, reusability, and so on—all characteristics that constitute a weak adhesive. Now we look to the broader invention family. What things, materials, or applications do we now use weak adhesive for? This question spurs a search for related applications that may also use weak adhesive action. An

Table 9.1   Post-it Note Adhesive Properties

Bonding effects
   Weak bonding or tacking
   Reversible bonding
   Reusable bonding
   Does not damage surfaces or leave residue
   Works with paper and other materials
   Promiscuous—without suitable substrata for binding it may attach to either
      surface that it is bonding together
Light controlling effects
   Resting state = cloudy
   Under pressure = crystal clear

answer is the fastening of paper and cloth. (We now have a list of related applications to work with and to compare and contrast with the possibilities of the Post-it adhesive.)

What other ways do we fasten paper together? Now the search is focused on specific inventions with weak fastening, such as those that bind paper. Thus paperclips, staples, and tape come to mind.

What are some of the disadvantages of paperclips, staples, and tape? Now we evaluate these other methods, looking for needed improvements. Our search reveals that paperclips, staples, and tape are likely to mar or damage the surface to which they are attached. (If we wish, ratings for the different fastening methods may be assigned, ranging from excellent to terrible. Staples and tape are particularly destructive of paper. Paperclips are limited to edge fastening, and cannot be written on.)

Can some of the specific functions of paperclips, staples, and tape be kidnapped or taken over by our new adhesive? (Remember, we are looking for a function for our adhesive, so alternative methods of fastening are examined in a search for potential new functions.)

Does the Post-it Note adhesive fulfill some of the functions of conventional paper fasteners and provide a better rating for some of the problems that we encountered? (A detailed evaluation is now made of competing inventions for fastening together things like paper or cloth.) We find that the Post-it adhesive is in fact reversible, does not leave residues, and bonds to a paper that can be written on.

These considerations in combination suggest that the Post-it adhesive applied to note-sized paper may have valuable functions, ranging from bookmarks to editorial comments to division markers. In short, we have found candidate purposes for the Post-it adhesive by going through a systematic set of questions, a set which is closely related to the categories that we introduced earlier for describing inventions.

The journey just taken through memory has been along an existing path, based on reconstructing known uses of the Post-it adhesive and on our knowledge of other book-marking methods. There is an advantage in focusing hindsight on a successful path, because the navigational steps on that memory path may be general and frequent enough to lead to new places in memory, where we may move along a new path in the same way.

Indeed, to test adequately the power of a hindsight heuristic, we must apply it to a *new problem* to see if it will generate an unanticipated solution. As an exercise in this art, we will ask the same questions about the other family of Post-it adhesive properties, those that involve light transmission. Examination of the bottom of Table 9.1 indicates those properties, a change in light transmission associated with pressure: the greater the pressure, the more clear the adhesive. All that I know about light transmission and the adhesive is revealed in the short property list at the bottom of Table 9.1. If 3M, or anyone else, has thought along this path, I am unaware of it.

Table 9.2 sets up a systematic set of questions and answers. It is essentially the New-Purpose Heuristic in action.

We note initially some interesting properties of the adhesive—how it responds to pressure, for example. Then we note an underlying dimension of variation, with light flow regulated by pressure. Next, we establish a broader invention family, light-controlling devices. But we need some specifics to work with, so we ask: What other things do we now use to regulate light flow? Our answer includes mirrors, curtains, and shutters.

A quick overall evaluation eliminates mirrors because they don't have similar light-controlling properties—they reflect instead of transmit light. We keep curtains and shutters because they do regulate light flow in a more similar way. A detailed evaluation reveals that curtains and shutters are all-or-none gates for light; any given point is covered or not. In contrast, the Post-it adhesive can continuously vary the intensity of light that passes through a point. That is interesting, and for some purposes it may be a big advantage over all-or-none curtains and shutters. This idea takes on new significance when we realize that the regulation of light also regulates temperature.

Certainly we may want to regulate temperature on a continuously variable basis. Is there some way to adapt the properties of the Post-it adhesive to better fulfill some functions of curtains or shutters? Yes, we can try to design a "solar window"—that is, a glass sandwich with a filling of Post-it adhesive. The window will be tied into an air compressor and a thermostat to control pressure, and indirectly the transmission of light and heat. On a cold day, the pressure will be turned up to make the Post-it liquid clear and allow sunlight to pass through. On a hot day the pressure will be dropped to make the Post-it liquid cloudy, so it can block the sunlight.

We have found a candidate application for the light-controlling

Table 9.2   The New Purpose Heuristic Applied to the Light-Transmission Properties of the Post-it Adhesive

General Question or Action	Specific Instance for Post-it Adhesive
List of known properties or parts?	Atmospheric pressure: cloudy High pressure: clear
Dimensions of variation?	Pressure, light flow
Invention family?	Light controlling devices
Other members of the invention family?	Mirror, curtains, shutters
Overall evaluation of match?	Eliminate mirrors; no obvious match Keep curtains and shutters; they also control light flow
Detailed evaluation of match?	Curtains and shutters are all-or-none gates for light; any given point is covered or not The Post-it adhesive allows continuously variable light transmission, so it might be better for some purposes than all-or-none devices
Adaptation needed?	Try to adapt Post-it adhesive for continuous control of light transmission One solution is a solar window, perhaps a glass sandwich with the Post-it adhesive as its inner core; a pump connected to a thermostat then varies pressures, and light flow is determined by the thermostat setting: too cold and a higher pressure is applied to clear the adhesive and let in more sun, and vice versa for too warm.
New purpose found?	Yes, at least a candidate one
Recheck?	See if this is really practical

properties of the adhesive. In short, we have found a new purpose. Of course, we do not know that our idea will work. We do not know what kind of pressure is required to change the light transmission properties; it may be excessive for such an application. Moreover, we do not know any cost, stability, or environmental impact considerations that will enter into having a solar window based on Post-it adhesive. However, we were able to think of the solar window application, which has surface plausibility, by applying the New-Purpose Heuristic. This suggests that the heuristic has both power and generality, beyond the context of bonding properties of adhesive, from which we abstracted it.

We can generalize further. In the solar window application, the continuously controlled variable of pressure on the liquid adhesive is used to regulate light flow and heat. Perhaps we can turn things around and use the properties of the Post-it adhesive to *measure pressure*. Suppose we need a pressure gauge in which a known light source intensity is on one side of an adhesive-containing slab, and the transmitted light is picked up by an illumination meter on the other side. Atmospheric or water pressure on the slab will then influence the amount of illumination picked up by the light sensor, providing a measure of that pressure impinging on the adhesive. We can generalize even further. Instead of talking about atmospheric or water pressure, we can talk about gas or fluid pressure. In doing so, we have followed a path from a few disembodied properties of a liquid adhesive to potential applications as diverse as solar windows and pressure gauges.

If we can find a way to make connections with new purposes, to invent new functions, we have found the least expensive of all ways for inventing products. What manufacturer of pantyhose thought of that product as ideal for keeping sand out of a tank's air intakes in the Iraqi desert? What manufacturer thought that a condom would be ideal for keeping sand out of a gun barrel? Can we go about finding new applications other than by good fortune? Can we do it systematically? I think so. The New-Purpose heuristic is just a first faltering step in that direction.

10

Multiple-Invention Heuristics: Joining

Another important form of heuristic that relates to multiple inventions is the *joining* together of previously existing inventions. Let's set the stage.

I'm sitting in a chair, making penciled notes for this essay. I've just made a mistake, so I turn the pencil in my hand and erase the error. What a convenient system. A pencil with lead at one end to make marks and an eraser at the other end to undo them. I wonder how someone came up with the idea of combining the previously separate writing and erasing ideas into one tool, the pencil with eraser?

One attempt to explain such joining of ideas is in Arthur Koestler's book *The Act of Creation*. Koestler uses the concept of *bisociation* to explain creative synthesis or joining. As he explains it, a bisociation is an unconscious connection between ideas. As such, bisociation is a concept that has a surface appeal, but closer examination reveals several problems. First, if we explain inventive thinking as the coming together of unconscious ideas, what should we do when the muse does not strike? Certainly, I do not wish to deny the role of unconscious processes, but if our sole way of generating inventive ideas depends on the unconscious, we must admit to little influence on the creative process and little possibility of teaching or learning about it.

A second problem of bisociation is that we know little of the unconscious principles used to bring ideas together. In fact, we likely know less about the unconscious than we know about creativity and invention. If this is true, it makes no sense to define the later in terms of the former.

A third problem is equally serious. In the absence of principles to explain the unconscious combinations of bisociation, we are forced to assume random combinations—and then subsequent selection from the most promising of those combinations. But random coming together leads to combinatorial or exponential explosion, the result if any idea or invention can be combined with any other. For example, suppose we generate new ideas by combining existing ideas or inventions pairwise. Then if the number of ideas or inventions is some large number N, the resulting number of pairs is $N \times N = N^2$, far more than the number of ideas in the mind or inventions in the world. The result of such random pairings may be as inspired as the claw hammer, but it will also include combinations like a dictionary and a bulldozer.

Actually, the situation is even worse than indicated by random pairwise matings because a given invention often can be the result of combining many other inventions. Instead of using pairs, if we randomly combine three inventions at a time then the number of possible triples is $N \times N \times N = N^3$. In addition to good ideas, we will have combinations like the dictionary-bulldozer-fishbowl, and undoubtedly the bad combinations will be far in excess of the good ones. The argument against combinatorial explosion is readily extended for combining four, five, or more inventions at a time, with complex inventions like cars and computers consisting of hundreds of separate inventions. Because combinatorial possibilities grow so rapidly, the result of random integration will soon lead to possible values as large as the number of electrons in the universe! If you don't believe it, compute the value of some large number with itself as an exponent.

We come now to a fourth problem of bisociation, that of selecting the meaningful combinations from all those we have generated. And even if we are only combining inventions pairwise, which of the many different ways of combining two given ideas or inventions should we use? For instance, the different functions of the Swiss Army Knife can be packaged much differently than they are. Instead of having a screwdriver at the end of a can-opener blade, might it have been incorporated with the knife blade?

These are difficult problems. To begin with, we need some terminology: When we combine separately existing inventions, and the separate functions of the component inventions are mostly preserved, the result is a *join*.

Types of Joining

I've been impressed with what can be done with an integrated tool like the Swiss Army Knife, and I want to make some of my own composite tools by combining other tools from around the house. Since I haven't done this before, I'm going to make it easy on myself by looking only at pairwise joins. For my first join I can put together a hammer and a flyswatter, next a coffee cup and a feather duster, and finally a plunger and a corkscrew. Somehow these just don't seem like good combinations. And Koestler's bisociation doesn't help very much, because my unconscious keeps putting together bad choices.

Even though it's easy to go wrong, clearly some joins are inspired combinations: the claw hammer, the pencil and eraser, a stereo receiver as the coming together of an amplifier and an FM tuner, and my Swiss Army Knife. What rules or heuristics can I try to increase my odds of producing useful combinations? That's an important question because there are so many ways of going wrong. We must find and use joining heuristics that will guide us, so we can avoid combinatorial explosion and the blind generation and dreary testing of all possible pairs.

Three heuristics come to mind. First, consider the joining of inverses. As previously noted, an inverse is an opposite or undoing invention. The example we started with is the pencil with an eraser on the end. The marks made by the pencil may be undone by the eraser. The claw hammer is another good example of a tool with a built-in inverse. The striker part drives in nails and the claw part removes them. The claw and striker are reciprocal inverses. The input for one serves as the output for the other, and vice versa. It is convenient to have a primary function and its inverse in the same tool. The combination means that we always have ready a way of dealing with error: Just apply the inverse. If one part of the tool is used for assembly, the other part of the same tool can be used for disassembly. These joint capabilities involve a minimum of switching time and effort. No longer must we put down the tool of primary function (the striker) to look for and pick up the tool of inverse function (the claw). All that is necessary to move from a function to its inverse is to change the edge of the tool and the grip used.

Said in another way, a tool with both a primary function and an inverse function has greater generality than either tool alone. The joined tool can take as its input nails in any state: not-yet-used-nails (pound them in) or used-and-inserted-nails (pull them out). The inputs and the outputs of the two tool functions feed into one another.

Joining a function and its inverse is likely to be advantageous. The join means reduced time and effort required to switch between the separate tools, the possibility of built-in error correction, the simultaneous capability of decomposition as well as composition, and increased generality of functioning. This suggests:

The Inverse Heuristic   Try joining those tools or devices that undo the actions of one another. These are often useful combinations.

I have phrased the inverse heuristic in the context of simple tools, because that makes it easy to understand. But the idea has much greater generality. As I sit in front of my word processor and simply back spaced over an error, it occurred to me how easy error correction has become: from erasers separate from pencils, to joined erasers and pencils; from typing and rubbing out and brushing, to white-out, to a special ribbon that allowed for the typed equivalent of a white-out; from computer line-editing commands, to now the simple movement of a key.

The idea of an inverse goes far beyond pencils and hammers. Another more jolting and complex example of an inverse is the emergency ejection system for fighter pilots. The pilot gets into the plane by his own relatively leisure means of climbing in. However, in an emergency, it simply won't do to climb out of the cockpit while the plane is going down in a death spiral. The solution is an explosive charge, activated by the pilot, that shears off the seat bolts and literally blows him out and clear of the plane. Inverses can take a number of forms, from simply stepping backward through our original actions to the evocation of very different principles like being blasted out of a cockpit. There is nothing mechanical about the thinking here. The inverse heuristic is merely a hunting license for finding ways of undoing.

We now come to the second heuristic to guide us in knowing what to join. We can get to it by generalizing on the idea of an inverse. Inverses involve *undoing actions* that are likely to occur *in the same context,* such as writing and erasing. Suppose we drop the requirement of undoing actions and leave in place the requirement of occurring in the same context. Then we have a definition of *complementaries,* two inventions used in the same setting.

Consider some examples of joining complementary forms. The carpenter's level is at once a straight edge and a level. It also has several orientations of bubble level joined together, one for vertical and another for horizontal layout. The same context, constructing a building, is likely to require all these functions, straight edge, vertical level, and horizontal level. The carpenter's level is a join of complementary inventions. As another example of a complementary join, consider the 35-mm camera with a built-in light meter in the same package. Instead of a separate camera and light meter, both used in the same context, the two are joined. On the chemical side, we have buffered aspirin, a combination of aspirin and an antacid, two ingredients frequently used together. Another chemical example is a combined shampoo and conditioner. Their combination makes sense because they are so frequently used together.

Successful joins of complementary inventions are those that provide increased functionality. The carpenter's level, the integrated camera,
and buffered aspirin are such joins. All of which suggests:

> *The Complement Heuristic*   Combine those tools or inventions that are used
> together in the same context. Do this especially if the separate inventions
> have complementary strengths and weaknesses, like the components of
> buffered aspirin.

There is still a third occasion for joining devices: when they share
properties or parts. Shared attributes and parts provide a basis for joining. This principle is apparent in the Swiss Army Knife. Its various
"blades" may share little more than the mutual requirements of having a common handle and a case. While the Swiss Army Knife is bor-
dering on baroque complexity, it is certainly successful in packing many
tools and functions into a small space. It does this by eliminating shared
properties of the individual tools, redundancies like separate handles
and cases. This suggests:

> *The Shared-Property Heuristic*   Whenever tools or devices share one or more
> attributes or parts, try to join them to eliminate redundancy, minimize size
> or space, or hold down overall cost.

For example, when two tools each have a handle, the resulting
join may need only one handle. Instead of having a separate computer
program for word processing, spread sheet functions, a data base, and
so on, build a single integrated program, with a common editor that
eliminates redundancy—and provides a consistent human interface. The
danger here is that the program will become a malignant form of the
Swiss Army Knife: some of everything that is uncomfortable to use.
Joining inventions based on an inverse, a complement, or a shared
attribute holds down the blast of combinatorial explosion. Let's look
now at some additional issues.

Parallelism and Emergent Functions

Imagine that you are walking at night with an armload of books and
managing an umbrella as well. Because it is dark, you keep stepping
into puddles and tripping over things; it is very awkward. If you didn't
have the books, you could hold a flashlight in one hand and the umbrella in the other. Now imagine an umbrella with a flashlight joined
with the handle; you can see ahead while holding the books with one
hand and keeping the umbrella overhead with the other. Remember
that just such an invention was suggested by a child inventor, Katie
Harding, in a *Weekly Reader* contest. Let's use it now as a more extended example.

Most joins of hand tools result in a *sequential integration:* one function at a time can be performed, even though the integrated tool allows for the separate functions of its parent tools. For example, the different blades of a pocket knife are used in a sequential fashion, one blade at a time; the A blade or the B blade, but not both at once. Yet with the joined umbrella-light it is easy to carry the books, be protected from the rain, and see ahead at the *same time*. When both flashlight and umbrella are integrated and *working simultaneously*, we have a join based on *parallelism*. We can see ahead with the light and keep dry while carrying an armload of books. Here the integration allows for a new function, something that we could not do with either tool separately: If we had one hand holding the umbrella and the other the light, we wouldn't have enough hands left to hold the books. Sometimes parallelism provides us with new capabilities, called *emergent functions*.

Parallelism in joins is so important that other examples are warranted. Two very early hominid tools are the rounded blade and the pointed awl of Figure 10.1. One way of getting to the pointed knife is to simply add the point feature to the rounded blade. Still, there is usually more than one path to a given invention, and another way is to form a join. Suppose the pointed blade is the join of a rounded blade and an awl. The two parent tools may then pass on their separate actions to the joined tool in several ways. In sequential mode, the blade part might be used for scraping, slicing, or cutting with a to-and-fro action, in the way of the earlier rounded blade. Or the point part might be used as an awl by applying a twisting forward motion, in the way of the parent awl. When used these ways, the functions are not new; the principal difference is the convenient integrated package of both functions joined in the same tool.

However, the blade and the point of the joined tool also may be used in parallel (simultaneously) for a new action—call it *etching*. Etching is appropriate for cutting a twisting groove in a complex design. It requires a new grip, with two hands on the knife simultaneously: one on the handle or upper back of the blade and the other on the lower back of the blade. One hand presses down on the point and the other pulls the blade toward the body while cutting the groove. This is a

Figure 10.1 Stone scrapper and stone borer, joined to produce a complex tool, the pointed blade.

simultaneous or parallel use of the join's components, point and edge. It is also an emergent function that is not possible with either of the parent tools: The awl point alone cannot cut out the groove, and the rounded blade alone cannot follow the twists of the groove, but the pointed knife as a joined tool can do both simultaneously.

When the elements of the join allow for simultaneous or parallel use, then new capabilities emerge: The joined tool is greater than the sum of its parents' functions. Etching a quickly turning grooved design was not possible with either of the parent tools, the rounded blade or the awl. Etching emerged only when the two components, blade and point, were used in simultaneous parallel operation.

The pointed knife demonstrates the characteristics of both sequential and parallel action; it is a very good join. To take advantage of the emergent functions that parallelism may offer, we often find it necessary to change the human interface and the grip. This was the case for the knife that was gripped with two hands while etching a groove.

All of this suggests an important principle:

> *The Emergent Function Heuristic*   When joining inventions, be on the watch for important new capabilities that are not present in either of the parent inventions. One such function is parallelism, the ability to use simultaneously the capabilities of each parent device.

Still more may be said about parallelism. In *simple parallelism*, we find multiple appearances of the same element or principle. This is exemplified by the saw; it has many teeth of the same form, and many of them cut at once. In fact, a one-toothed saw is a ridiculous idea. Another example is a rasp or file. Again many teeth or grooves are in simultaneous operation. Still another example is many people painting a wall at the same time, and performing the same operations.

In *compound parallelism*, we find different elements or operations concurrently at work. The pointed blade and awl is one example. Another example of compound parallelism is a nineteenth-century American frontier favorite, an apple-peeling device resembling a lathe that simultaneously attacks apples by peeling, coring, and slicing them. These seemingly simple devices are beautiful examples of compound parallelism in which several different operations are performed at once as numerous cutting surfaces impinge on different parts of the apple. Still another example is the video cassette. It is at once a set of reels that are joined together in an encapsulated unit that functions as a reel, a cartridge, and a protective case. Finally, the umbrella/light we began with is a conceptual example of compound parallelism at work.

Levels of Joining

Even when we have heuristics for helping determine which two of many inventions to join, that is just the beginning. Two inventions can be joined to one another at a variety of levels. The levels of joining are loosely analogous to different forms and strengths of chemical bonding. Although we have no metric for the strength of an invention join, we can clearly order joins according to the degree of integration—at least for some domains such as hand tools.

To provide focus, let's concentrate on joining simple hand tools, those without internal parts that move during tool use. The joins of simple hand tools vary from weakly integrated to more strongly integrated. The level of a join does not mean "better" or "worse." That's a separate issue. The meaning of *joining level* is best explained by example, ranging from weak to strong integration.

An *unrelated assemblage* is the weakest form of integration, a null or zero join. It is just what the phrase indicates, an assemblage of unrelated tools and other items. A junk drawer full of miscellaneous items serves as an example. Some people will worry about calling this non-combination a join. But it is useful to consider limiting cases in a classification, just as it is useful to consider zero a number.

A *related assemblage* is a higher level of integration or join, but the separate inventions are still without any physical connection to one another; it is an assemblage of related tools that are separate from one another. An example is a set of wood carving knives or perhaps a tool kit that has a rational design behind it, one that allows for many different tool functions that build on and complement one another.

An *attachment join* is a much stronger form of integration, with a physical—although transitory—connection between components. Typically, a central unit has specialized attachments that fit with it in a modular way. Examples include a vacuum cleaner and its attachments and an oxy-acetylene torch with separate welding and cutting heads that may be alternately attached and removed. If the attachments are enclosed in the main unit or case when not in use, we may refer to it as a *case join*, a close relative of the attachment join.

A *loosely bonded join* is a still tighter form of integration. It occurs with the separate blades of a pocket knife. These blades have a common handle to which they are physically and permanently attached; there is an enduring physical connection. However, selection of a function requires some setting or "dialing" to get out the right blade. Other examples include a key ring and a mode switch on a digital watch with a host of different capabilities. In each case, some effort is needed to set the right function.

In a *moderately bonded join*, the separate functions are on the same blade. Typically, those functions occur on different edges of the blade. Hence in the Swiss Army Knife one blade has both a can-opener edge and a screwdriver edge. We can open a can or drive a screw simply by using the different edges of the same blade.

An *adjustable join* is in a sense the combination of *many tools* of different sizes in the same package. An example is the crescent wrench, which may be regarded as the join of many wrenches of different fixed sizes.

A *tightly bonded join* is the strongest form of integration for hand tools. It occurs when the same tool edge has more than one function, and the exercise of those functions is simply a matter of the actions provided by the user. For example, the blade of a shovel is used for the functions of digging, scooping, and chopping, where function is determined by the hand or foot grip on the shovel and the actions of the body on the tool. If those functions had their origins in different digging tools, then the shovel is a tightly bonded join.

The joining classification just developed makes sense after the fact. How can we validate such a scale? What would it mean to do so? To start with, we can try to systematically vary the level of join when combining arbitrarily selected tools. If we can do so, greater confidence is attained in the generality of the scale. The prospect of joining two tools at different levels of integration may lead to a new and more systematic concept of design. The worthiness of the resulting inventions must still be determined, but that is a different issue.

To determine if the concept of joining level has generality beyond the specific examples cited, I want to consider the unlikely case of a hammer and a screwdriver joined at different levels of integration. These combinations are not being extolled for their usefulness as new tools. Instead, the intention is to see if we can extend the concept of join to *new tool combinations*. The outcomes of a thought experiment—examining different levels of join—are illustrated for the hammer and the screwdriver in Figure 10.2.

First, we can readily see a separate, unrelated assemblage of screwdriver and hammer. Perhaps both happen to be in a junk drawer that has no obvious organization. When the same screwdriver and hammer are both *intentionally* in the same tool kit, they constitute a related assemblage.

Moving on, a screwdriver blade and a hammer head may be attached to a common handle, as in Figure 10.2, to form an attachment join. The screwdriver blade and the hammer head are separate attachments to the handle. This is not an elegant tool, but the combination has been commercially marketed. Several screwdriver blades are stored in the common handle and can replace the hammer head. Attachment joins often require extended setup time to put in place the right attachment; they also require some way of keeping track of the attachments that are not in use.

When the screwdriver and the hammer are spatially combined to require a dialing or setup, we have a loosely bonded join. The invention illustrated is certainly crude. No indication is shown of how to keep the blade and hammer head in a rigid angle with respect to one another; and the screwdriver blade does not make a satisfactory ham-

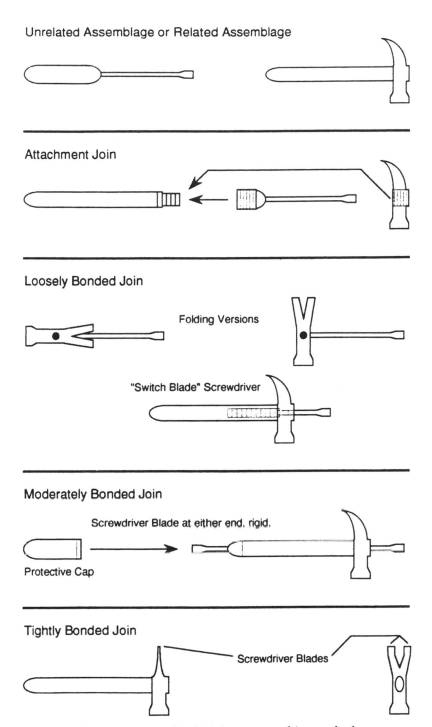

Unrelated Assemblage or Related Assemblage

Attachment Join

Loosely Bonded Join

Folding Versions

"Switch Blade" Screwdriver

Moderately Bonded Join

Screwdriver Blade at either end, rigid.

Protective Cap

Tightly Bonded Join

Screwdriver Blades

Figure 10.2 Levels of joining, a screwdriver and a hammer.

mer handle. Yet fine-tuning can help these problems. For example, making the screwdriver shaft much thicker, except at the tip, provides a better hammer handle. However, no one is likely to get rich marketing this tool. Another variant of a loosely bonded join is a "switch-blade screwdriver." The screwdriver blade is encased in the handle until a button is pressed to release the spring-bound blade. Although setup time is again required prior to use, tools with loosely bonded joins are likely to require much less setup than attachment joins.

A permanent form of screwdriver protruding from either end of the hammer is shown in Figure 10.2, and it constitutes a moderately bonded join. Unlike the switch-blade screwdriver, dialing or extruding a blade is no longer required. Having a screwdriver blade protrude from either end of the hammer has at least one desirable feature, rotating the hammer head produces tremendous leverage on the screw. If the screwdriver is at the handle end, a disadvantage also presents itself, the possibility of stabbing one's arm with the blade while hammering. We can solve this problem by covering the screwdriver blade with a protective cap when hammering, but this does not increase functionality, only safety. If the screwdriver blade sticks out the other end, past the hammer head, there is another disadvantage: hammering near a wall is impossible. With a moderately bonded join, we see that a change in the tool used involves a change in both the tool edge and the grip on the tool. However, the setup time is typically much quicker than for lower levels of join.

Apparently, the hammer and the screwdriver do not fit the category of an adjustment join. However, a tightly bonded join may be possible. If we slightly straighten the hammer claw, as in Figure 10.2, we can use its tips not only to pull nails but also as screwdriver blades. This duality of function for a given tool edge defines a tightly bonded join. When claw tips are used as screwdriver blades, a bonus is had: Each of the several claws makes possible a different size of screwdriver blade, all on the same tool. This particular join of claw hammer and screwdriver may actually be useful. Notice that the setup time for a tightly bonded join is minimal; simply change the grip or action applied to the given tool edge.

Developing this classification for levels of joining is informative, but we might wonder if it is complete. I certainly don't think so. Other more complex examples of adjustment come to mind. In some of these cases, the parts change dynamically through *automatic action*. A 35-mm camera in which the lens openings and shutter speeds are automatically controlled serves as an example. So the level of joining idea can be expanded further to include inventions like the camera's automatic control: we simply point and shoot; everything else is handled for us.

Of course, when domains other than simple tools are considered, very different principles may apply. Indeed, the most sophisticated areas of invention, domains like electronics and chemistry, are built largely on a theory and practice of discovering and formulating principles of joinery or combination in much more detail than those applying to tools. Nonetheless, I find it revelatory that a domain as simple as hand tools seems to have a great deal of structure in the way its different forms are combined, but that structure is largely undescribed. Clearly, one of the first things to do in an invention domain is to try to figure out the often unspecified principles of joining elements.

Let's finish this chapter by coming back to Katie Harding's conceptual invention, the umbrella flashlight. It is a join based on complementarity—often we need to use an umbrella and a flashlight at the same time. It is also an emergent invention because we can carry an armload of things while using both the umbrella and the flashlight, something that we can't do with the unjoined ingredients. The umbrella flashlight is also a join based on compound parallelism because the separate ingredients have quite different functions. And finally the level of joining offers several possibilities. It could be an attachment join, if the flashlight can be inserted and removed from the handle. If the light is built into the handle, it is probably a moderately bonded join. But if the angle of the light can be varied with respect to the handle, then it is also an adjustable join. Deep complexity can underlie surface simplicity, and the inventions of children can help us see that complexity.

11

Transformational Heuristics

My tool kit is spread out in front of me: awl, knife, chisel, saw, file, rasp, among other rusty tools. History tells me that these tools arose at very different times for undoubtedly quite diverse purposes. Nonetheless, my interest in invention has taught me that different forms often exhibit striking similarities. I'm wondering if that might be true for the devices in my tool kit also?

Let's now engage in some freewheeling mental play in the form of thought experiments that attempt to organize hand tools. The underlying element of organization, the tooth, does its work by acting as a wedge; and as the tooth-wedge is transformed, it gives rise to a family of tools.

Transforming an Element

Just as interlocking strands of DNA can magically transform themselves into a person, so too some inventions can change into others, by *transforming an element*. This is almost certainly not the way hand tools or most other inventions have been generated in the past. Instead, I want to offer an after-the-fact rationalization and systematization. Mathematicians do this all the time to bring order after-the-fact to their systems. This notion of a thought experiment is a time-honored one in science also, and it was one of Einstein's standard methods of proceeding. Even if our ordering of inventions is after the fact, we can use the transforming principles to show connections between tools, and perhaps we can discover how to invent new forms in the future, to change some caterpillar into a moth, if not a butterfly. Let's work into the transformation idea by way of the knife.

The knife is based on the principle of the wedge. Forces are applied over a broad range (by shoulder, arm, wrist, and hand) and converge on a very narrow wedge-shaped blade or point region. The result is a tremendous amplification of pressure, the force per unit area. The greatest amplification of pressure is where all the forces are focused on a point or a physical form like a tooth. In fact, I will show that the fundamental element that underlies the knife can be thought of as a tooth. With these ideas in mind, we can now apply different spatial transformations to the tooth to generate a variety of other tools.

The awl is the most simple example of a tooth-shaped tool. It is used to punch or bore holes. It is essentially the tooth of Figure 11.1, with a handle. To operate the awl, simply push and rotate with the wrist, punching a hole in leather, wood, or some other material.

The saw of Figure 11.1 can be thought of as copying and alignment transformations of a tooth. The tooth is repeated many times and the copies are punctuated along a line, leaving a space between each tooth. Because the teeth fall on a straight line, this is a one-dimensional transformation of the tooth. Our thought experiment is telling us that the important idea behind a saw can be realized by transforming the tooth.

Next, let's look at the blade and the pointed knife, as indicated in Figure 11.1. While the saw resulted from punctuating the tooth along a line—alternating between teeth and space—the blade can be thought of as densely or compactly packed teeth along a straight line, without spatial gaps. When this is done, a continuous blade edge results. The smooth blade edge allows new modes of action, separate from those of the tooth. Examples of these actions include scrapping, whittling, or slicing. This repetition and dense packing of the tooth along a straight line produces a simple rectangular blade, one without a point. However, if the tooth is turned upward toward the end of the blade, we get a pointed blade. Because of that slight upward turn away from the straight line, the pointed blade may be regarded as almost a two-dimensional transformation; call it one-and-one-half dimensions.

What if a set of teeth are arrayed in a full, obvious two dimensions? If the teeth are relatively large and separate, then a rasplike tool results, as in Figure 11.1. If the teeth are small, densely packed in no special pattern, then a grinding or sanding surface results, with the cutting done by the sharp edges of many individual grains.

The file is still another transformation of the tooth. Instead of starting with the tooth, use one of the higher order elements we have already generated, the blade. If a blade is made very small and then

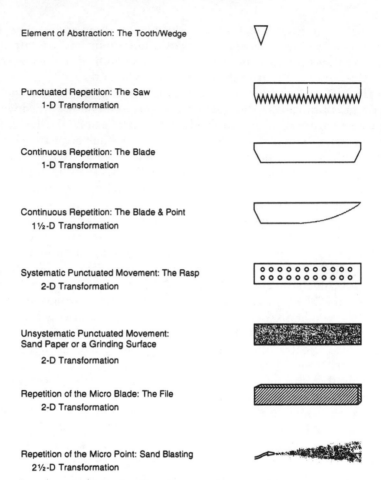

Element of Abstraction: The Tooth/Wedge

Punctuated Repetition: The Saw
  1-D Transformation

Continuous Repetition: The Blade
  1-D Transformation

Continuous Repetition: The Blade & Point
  1½-D Transformation

Systematic Punctuated Movement: The Rasp
  2-D Transformation

Unsystematic Punctuated Movement:
Sand Paper or a Grinding Surface
  2-D Transformation

Repetition of the Micro Blade: The File
  2-D Transformation

Repetition of the Micro Point: Sand Blasting
  2½-D Transformation

Figure 11.1   A tooth and some of its transformations. In our thought experiment, this element of abstraction, the tooth, serves as the building block for important tools that are generated by different spatial transformations.

repeated many times at the correct angle, a file results from the com-bination of microblades. The blade involves more spatial organization than a random array of grinding teeth. The microblades are at a fixed distance and orientation from one another, and they are at angles with respect to the work to let waste cuttings escape. The actions of the microblades are in parallel, each blade doing its own work. The mi-croblade array is two-dimensional.

Finally, we come to a tool that is not in my kit but can also be generated from the tooth. The generation requires a dimensionality greater than two. So far we have assumed that the tooth is fastened to a surface. But if we were to allow the tooth to decrease in size to a microscale and were then to "throw" many of them at a surface, we would have the equivalent of sand blasting, a technique that allows us to cut in the picture plane in front of us and also in the depth plane extending away from us, a multi-dimensional capability. The underlying reality, of course, is a swarm of sand grains with sharp wedgelike edges, no one of which cuts very much, but which collectively can polish and cut with surprising speed. While sand blasting operates in both the picture plane and the depth plane, it is not as fully three-dimensional as electroplating or an acid bath, each of which works on all sides of an object at once. For these reasons, I think of sand blasting as working in only two-and-one-half dimensions.

In summary, the knife and many other tools can be viewed as variations of the tooth—and its underlying principle, the wedge. All these tools result from a small group of spatial transformations: copying a tooth many times, and moving the copies along a line, or a plane, or in a third dimension. Many tools created by these transformations also are blades or blade edges in complex hand tools with multiple joins, such as the Swiss Army Knife.

What we have done here is both simple and powerful. An *abstraction heuristic* was used to arrive at the tooth or sharp point as the building block for the knife. Then we applied *repeating and packing transformations* to the tooth, arranging different configurations in space. But more than this, we *generalized dimensionality* and moved from one to two to three dimensions—all the while changing the tooth's scale. A small number of manipulations like these, when applied to the tooth, generated a number of standard tools.

Claims like these raise a critical question: Are we not just taking a geometric abstraction, the point, and showing that it can generate any structure in three-dimensional space, a not so original idea? No, the tooth is a real three-dimensional form that operates on a real physical principle, the wedge. It is altogether different from an abstraction like a geometric point—which after all has no size or shape, only location. Nor have we created some abstraction that explains everything, something so ethereal that it really explains nothing. Spatial transformations of the tooth are not sufficient to produce all hand tools. For example, a bottle opener does not operate on the principle of the tooth or wedge; instead it is an example of the lever. Not even all cutting tools operate on the wedge principle. Thus scissor "blades" are really levers with a common fulcrum, and they work through a shearing rather than a cutting action.

Another comment is in order. As appealing as this exercise has been, general heuristic principles of invention must yield to specific problems. The idea of a saw is more than a line of punctuated teeth. Serious cutting quickly reveals that such a saw binds. Additional refinements are required to free the saw from binding and make it a viable cutting tool. The teeth must be staggered or "set." To make an invention work well, very delicate and precise relations are often needed to fine-tune a basic idea. That is certainly true of transformations to create tools.

What then is the status of invention by transformation? What we have done here is to show that an abstraction, the tooth, can be spatially transformed in a variety of ways to yield a large number of hand tools. At a minimum, we have produced a powerful organizing scheme. But it may be something more as well. In fact, an acid test of its efficacy is to apply spatial transformations to the tooth that will produce new and useful tools, ones not now in existence. I cannot presently do this. Instead I'll simply rest on the claim that the transformation approach brings order and sense to a family of inventions. This is not a bad perching point. The chemist's periodic table had a similar beginning.

Transformations of an Egg

In the mental path we just took, we played with an element of abstraction, the tooth, and applied different spatial transformations to generate different tools. How general is this approach? Perhaps abstract elements and transformative relations are specific to hand tools? Perhaps transformative invention is not applicable anywhere else? To answer the doubts these questions pose, I want to show that elements of abstraction and invention by transformation have generality beyond simple hand tools. At the same time, I want to show you that an invention path can be traveled in different ways, sometimes starting at one point, and other times from another point—yet, all these directions wind together.

I'm looking at an egg. I'm fascinated with its form: It's austere and beautiful, incomparable architecturally. I wonder what truths of invention are hidden here.

What is common to an egg, an arch, a dam, and a dome? Can we find spatial transformations that link them? Why are hen's eggs not spherical? These are questions to consider as we extend the transformation story to architectural form. Let's begin with the egg and then abstract from it as we slice it, twist it, push it, and stretch it to see what else we can create.

It is time for a "wet lab." To protect your hand and to apply forces evenly, wrap a towel around an egg and squeeze it very slowly. Gradually increase the force until the egg breaks, or until you give up without breaking it. Whatever the result, you may be surprised at how strong an egg is when the forces are distributed on its surface and gradually applied.

Did the exercise of squeezing the egg give you any ideas? Is an egg stronger along one axis rather than the another? Get out another egg and squeeze it along a different axis. In the egg you have discovered a geometric form that is at once strong and light (when it is empty). How could that shape be transformed in other ways? If the size of the egg is scaled up to dwelling size, it provides an interesting architectural structure, but the rounded "floor" of the scaled up egg is a problem for a living space because every time you enter the structure, you slide toward the bottom center; there is no flat surface to stretch out on. This problem is solved if the expanded egg is first sliced in half, in the manner of a soft-boiled egg, and the base is placed flat on the ground. That way the half-egg acts as a shelter over a flat floor. What should such a structure be called? Perhaps a *dome?*

The dome's architectural form is interesting, but you don't know how to build one out of bricks. Maybe, if the dome were further simplified you could build it with bricks? Two closely adjacent slices through the upright dome will transform the three-dimensional dome into a two-dimensional form. For lack of a better name, call the result an *arch.* What can the arch be used for? You play with the arch form, and you find that it is not strong when pushed from the side, but when pushed down from above, it is very strong. That suggests a substitute for the way you currently build doorways and windows. Usually you put a straight wooden beam over the top of an opening, and then you continue to lay bricks on top of the beam. The problem is that after a few years, the wooden beam rots out, and everything comes tumbling down. Moreover, trees big enough for a large beam are becoming ever scarcer. By building this arch form out of bricks, you will solve both of these problems. You think there is a way. To keep them from falling down before you get them all laid out as an arch, you need a temporary curved wooden or earthen support, but that's easy enough. You try it. What you've done looks very good. In fact, it's an elegant aesthetic form, a bonus. You may guess that baring major shifts in the earth, the arch structure will last a long time because the weight above it, unlike that over a straight beam, is channeled to its sides.

Now that you have created an arch by slicing the dome, which was a slice through the egg, you wonder what other useful forms you can find by further spatial transformation. The answer is not long in coming. Recently, in your town, torrential rains resulted in flooding that tore apart the dam you and others had worked so hard to build. The dam structure cut straight across the stream, just like the straight beams over the doors and windows. You wonder if the arch form might also be used to build a dam. What is required is laying the arch on its side, with the peak pointing upstream. Then the water pressing against the dam will be pressing against an arch that has been laid out on its side, far stronger than a straight dam.

You reflect back on what you've done. The egg had the interesting property of resisting stress when squeezed or pressed. You then progressively abstracted simpler forms from the three-dimensional egg, again looking for stress resistance.

Now let's reverse directions on our invention path, a common strategy. This time we start with the arch, once more transforming it, and then we'll see where we end.

As we have noted, an arch is a form that affords a strong way of holding up wall masonry over a door or window opening. It is stronger than a horizontal beam because it spreads the downward forces to the sides of the opening. But it is harder to build than an opening spanned with a horizontal beam. The arch is vertically oriented, with the peak at the top and the base at the bottom. It is a wonderful form, strength clothed in grace.

Now that you have the arch, what can you do with it? A common spatial transformation is rotation. Suppose you rotate the arch by laying it on its side so it can resist horizontal forces. Now you need to find an application, and you ask yourself, what forces act horizontally? Water spreading out, as in a flood. Therefore, a horizontally oriented arch could function as a strong barrier to water, what you now call a dam. The arch form might work as a substitute for your previous method of holding water back, the straight dam, which had problems similar to those of a straight beam over a door. However, the straight dam has some advantages over the arch-dam: it uses less material (a straight line is the shortest distance between two points), and it is easier to build. But for strength and aesthetics, there is nothing like a gently curving arched dam. It may be just the thing to hold back the next flood.

Another path also leads from the arch to the dam, but by a mental connection or link instead of a spatial transformation. Since the arch has shown itself much stronger than a horizontal beam, what other straight-beam structures can you think of that must withstand forces? A straight-wall dam. So why not replace it with an arch-like dam? Here of course the link is through "other straight-beam structures," and only after the link has been forged do you rotate the arch.

Still other invention paths lead to the same destination. For example, suppose you just had a straight-wall dam burst. What can you do to make the new dam stronger? Make it thicker—but that will require more work and materials. What next? Well, do we build any other structures that are stronger than straight barriers at right angles to forces? Yes, the arches above doors and windows, so why not rotate an arch ninety degrees into a position to hold back water? Spatial transformations can operate in both directions. Depending on the direction, they take us from form to need or from need to form. Whichever way we go, we are not finished. Other more complex variations on the arch theme whisper to us.

Now let's pick up the main thread again. What would happen if we stretch the arch and change its size scale? As we face an arch (Figure 11.2A), if it is stretched into the third dimension or depth plane, the result is another architectural form, the corridor (Figure 11.2B). Or if the size scale of the corridor is greatly extended, the result is a nave structure characteristic of the great cathedrals. The nave is typically a long open space with arched ceiling and without inside vertical supporting beams.

Now suppose that you take one of these corridors with an arched ceiling, make it very small in scale, and copy or repeat it many times to produce many little corridors. What is the result? With a few other variations, corrugated cardboard can be produced. A cross section of cardboard is shown in Figure 11.2C. Transformations for getting to it from the arch include the following:

- Decrease the scale of the arch.

- Stretch the arch into a corridorlike structure.

- Repeat the corridor structure many times, as in Figure 11.2B. This will provide support strength over a surface area.

A. The basic arch.

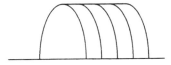

B. The arch repeated to form a corridor.

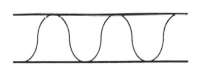

C. Cross section through a corrogated cardboard box. A series of corridor structures repeated in an alternating mirror image manner. A holding and protective veneer is glued to each side.

D. Aerial view of an arch rotated about its vertical axis to produce a dome.

Figure 11.2　Some spatial transformations of the arch (A). Stretching produces a tunnel or corridor (B), repeating and alternately flipping the arch yields the internal structure of corrogated cardboard (C), and rotating about a vertical axis produces a dome (D).

- Alternate the orientations of the corridor peaks, up and down, as in Figure 11.2C. This will provide resistance to forces coming from either side.
- Add a facing veneer to both sides to keep the result from buckling, as in Figure 11.2C.

The result of applying these transformations is corrugated cardboard. Altogether, an interesting journey from an abstract form, the arch, to the cardboard box that is both light and strong. As we will see later, the heuristics underlying the invention path from arch to cardboard are remarkably similar to those that will carry a hook and eye structure into Velcro. Is this just a coincidence, or should we take this heuristic path more often?

Still other variations on the arch remain. The simple arch is a two-dimensional form. To be sure, the walls of real arches already have some thickness, but that is just the accommodation of abstract form to a physical world in which building materials have thickness. What else can be done as you move beyond two dimensionality?

Try to rotate the arch about a central vertical axis. What is the result? A new architectural form that is called a *dome*, as shown in Figure 11.2D. All that remains is to find out how to build it. You want to do so because it suggests the possibility of a huge expansive area without any distracting vertical pillars. Perhaps the dome would be just the structure needed for the construction of our new cathedral?

You can now copy the dome and place the two domes base to base. The result is a form that encloses space. Very strong and light. You choose to call this an *egg*. What we learn from these mental rotations is that inventions can move along multiple paths, or in different directions along the same path.

But how did you get to that beginning, the egg? Why do chickens cloth their embryos in egglike forms? Suppose you were Nature's Chicken Designer. Why the egg form? Well, it has to be rounded for at least two reasons. No chicken will consent to pass an egg with sharp edges, and therefore evolution will not allow it. Another reason for the egg shape is its strength. All right, but why an *ovoid* shape instead of a *sphere?* Why not have your chicken lay spherical eggs? Frogs, fish, and the like do. Spherical eggs can withstand forces from all sides equally well. True, but the egg probably does not have forces equally imposed from all sides. The maximum forces are probably at the ends, at one end the muscles pushing to pass it and the force of impact at the other end.

Still another reason for the ovoid egg form is the need to get a lot of embryonic material in a container of small diameter, so the hen will be able to pass it. That also suggests the elongated ovoid form. As Nature's Chicken Designer, you will allow evolutionary forces to work more rapidly in elongating the egg shape than in expanding the diameter of the duct through which the egg passes. The latter is a more difficult design problem, because it entails changing many dimensions of the chicken.

You are now at the egg farm where individual eggs are being packed into the dome-shaped compartments of one-dozen egg-containers; these containers are then being packed into corrugated cardboard boxes built on a transformation of the arch. All this suggests that invention may start with an element such as the arch or the egg. Then transformations may be applied systematically to that element to find where they will lead. If you do this as a computer, you will find that these simple spatial transformation heuristics are too vague to be helpful. If you do this as a human, you may find that an egg suggests many architectural forms. If you do this as a god, you may find that the starting point of an arch suggests a good way to design a chicken egg.

A last word here. Sometimes when I have described how hard it is to break a towel-wrapped egg by slowly squeezing it, people will say the difficulty is due to the relative incompressibility of the liquid material in the egg, rather than its shape. That's not true. Find a Leggs' Egg, a packaging container for women's pantyhose. This egg comes apart in the middle, essentially forming two dome structures. Start squeezing one of the domes from various directions. You will find that when squeezed around the middle the dome is easily bent out of shape. But if you place it on a flat surface and push down on it from above, it is very strong and rigid. Now if you put the two domes together to form a full egg, you will find the structure resists deformation from whatever direction you apply a squeezing force. All of which is obvious to the mechanical engineer, based on theory. The point is that people employed arches and domes long before there was any adequate supporting theory. And they will do so in the future for other invention problems.

12

Discovering
Heuristics

Last week I bought a new set of tires for my car. I had to wait for some time for the tires to be put on, and, during my wait, I wandered through the counters and display of the sewing section of a nearby store. These experiences prompted me to think about two seemingly unrelated ideas: What are the origins of this device that moves us around so readily, this device we call the wheel? And where did the idea for sewing come from?

Let's use these questions as a vehicle for how to find heuristics, since by now you're probably wondering where heuristics come from. The answer is simple enough. Usually, heuristics are tentative generalizations abstracted from actual cases of invention, and ideally they are tested for generality by seeing if they have explanatory power for other inventions.

Reinventing the Wheel

The cart wheel is one of history's great inventions. It makes possible the movement of heavy loads with minimal human effort. Our purpose here is to discover and evaluate heuristics that may have guided its development. The idea is that the movement to such an important invention must embody important heuristics, and we need to learn them. Several competing accounts of the wheeled cart's origins will be played out, and each suggests interesting transition heuristics. Once more, I'm not making strong claims for actual historical development. Instead, for each proposed origin, I want to find transition principles that are both simple and powerful, with the underlying assumption that the transitions follow the operations or tracks on which the mind most readily moves. The tracks may then be examined for heuristics that underlie them, with the goal of providing economical and generalizable principles for inventing.

The wheeled cart made its appearance about 3500 B.C. in the Middle East. If we were giving Nobel Prizes for early technology, this invention would surely warrant one. A simple two-wheeled form is shown in Figure 12.1. At least three ancestral inventions have been suggested for its origin: the potter's wheel, the land sledge, and a platform on rolling logs.

As our first story of origin, let's consider the transition from potter's wheel to the two-wheeled cart. In this account, the path of invention *starts with the wheel itself* and then adds a vehicle function in the form of a cart. The closest thing to a wheel itself is a potter's wheel, which developed about the same time or earlier. One form of the potter's wheel is shown in Figure 12.2A. The wheel sits on top of an axle-like column, but the column does not penetrate the wheel. Any potter's wheel of this form must have been moved from time to time. The easiest and most natural way to move it is to flip it to its edge and then roll it. When this is done by hand, it is apparent such rotation produces *low friction* in comparison to picking it up or sliding it.

The next step is to forge a connection or link with an existing

Wheeled Cart

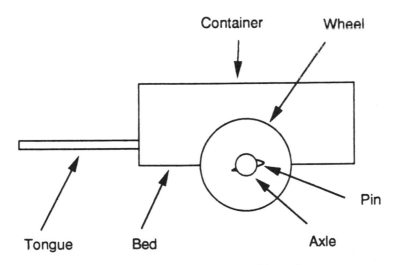

Figure 12.1 The wheeled cart and its major components.

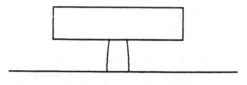

A. The potter's wheel (side view)

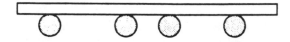

B. Platform and logs (side view)

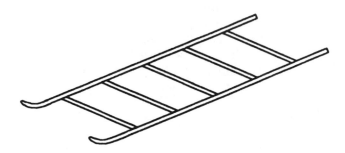

C. The land sledge.

Figure 12.2   Possible precursors to the wheeled cart: the potter's wheel, the platform and logs, and the land sledge.

sledge or platform used to transport loads. A sledge is essentially a land sled, and a platform is a kind of pallet to put goods on so they can be moved as a group. When a sledge or platform is moved, substantial *friction* is apparent. All of this suggests another heuristic, *a join based on complementarity*. Likely components are the wheel and sledge. The potter's wheel has low rolling friction, but no carrying capacity when vertically oriented. The sledge or platform has large carrying capacity, but high friction when moved. Putting the two together produces the best of both worlds, load carrying ability with low friction. But this is only a conceptual join so far. Specific steps are still needed in the transition from potter's wheel to cart wheel to make a workable combination. These steps include:

- Flipping the potter's wheel 90 degrees (this must have been done already when moving it).
- Adding an axle to the wheel (the potter's wheel already had a rudimentary axle, as shown by the supporting column in Figure 12.2).
- Poking the column axle through the center of the wheel, after flipping it to a vertical position (the axle may have naturally worn through the potter's wheel anyway).
- Pegging the axle to keep the wheel from falling off.
- Adding a second wheel at the other end of the axle, to provide stability. (A repetitive or copying operation to introduce a second wheel for stability, much as the second "backbone" was introduced in an earlier chapter for the transition from pack animal to chassis.)
- Joining the wheel and axle to an existing vehicle (platform, sledge, raft).

Notice that getting the wheel-axle configuration and the cart connected is an interfacing problem. While the details of going from the potter's wheel to the two-wheeled cart are both numerous and momentous, they make sense if one already has the idea of a complement: combine an easy moving form that has no load-bearing capability (the wheel) with a form that can carry loads even though it generates a lot of friction (the sledge or the platform).

Now let's consider a second story of origin, a transition from the platform-on-logs to the wheeled cart. Figure 12.2B shows a combination of a platform and rolling logs. The emphasis here is on *starting with a vehicle* and trying to decrease its friction by placing rollers, rocks, and so on, under it. A rudimentary vehicle such as a platform can hold a heavy load but it encounters considerable friction in its movement. The friction may be decreased on some terrains, those that are both solid and smooth. Even easier for movement are those terrains with round stones or logs in the right orientation. Under this scenario, the fortunes of chance enter the mind prepared for a method easier than pushing and pulling a platform. It is a kind of hill-climbing search that is more than metaphorical. The steps may have been as follows:

- Pushing the platform and cursing the effort.
- Noticing that the platform moves more readily when it goes over a small rolling log.
- Deliberately placing a log underneath.

- Placing more than one log underneath; trying several. (Notice a repetition principle is at work here.)

- Bringing a previously passed-over log from behind to the front or leading edge of the platform, as the platform begins to outrun the logs. (Another repetition begins here.)

- Inserting the log again under the platform.

- As often as necessary, retrieving logs from behind and moving them to the front. (The entire repetition or loop begins again.)

In this account, the log doubles as quasi-axle and wheel, floating back in position under the platform as it turns. Also, what one log does well in decreasing friction, several logs do better. Even with several logs, the platform will soon pass over them all, and they will have to be reinserted under the front edge. Each of these problems can be corrected by a profound act of parsing and abstraction—by seeing the separability of the axle and wheel functions, a fixed axis around which something rotates.

Once the log is seen as having both wheel and axle functions, the problems discerned can be corrected by fixing the rotating log in position. Simply keep the axle in place and allow the wheel to turn about it. This may be done by dividing the functions of axle and wheel. The axle can now be attached to the vehicle, while a separate wheel turns around it. Said in another way, the axle interfaces between two complementary parts, a minimum-friction wheel with no load-carrying capability and a high friction vehicle with extensive load-carrying capability. Once again, this is a join with perfect complementarity.

Finally, let's consider a third story of origin, the transition from sledge to wheeled vehicle. Another way of starting with a vehicle and reducing the friction under the platform part is to fix in place some runners under it. Friction is lessened by having a smaller surface area in contact with the ground, but this approach has its limits. When the surface area becomes too small, the vehicle will sink into mud or dust. A useful geometry for attaining the low friction of a small surface area, and for also avoiding the problem of sinking into soft ground, is a long runner like that on a sledge, a vehicle that is essentially a land sled. Figure 12.2C shows a sledge. Any irregularities or soft spots in the soil are likely to be cancelled out by harder spots at other points along the runner. The result is minimal contact and friction with the terrain, while also avoiding the problem of sinking into soft spots. This is still another case of complementarity at work. The invention steps may have been as follows:

- Noticing that pushing or pulling a loaded vehicle in a direction parallel to the long axis of the fixed-logs of a platform is easier than pulling across the wide axis.

- Deliberately putting a skid under the platform by generalizing on the idea of a long axis.

- Adding a second skid under the platform for stability. (Again, the idea of repetition is at work, like the earlier chassis example.)

- Instead of moving the skids to accommodate the changing position of the vehicle, attaching the skids to the vehicle, along the long axis.

What we have now is a relatively friction-free land sled. To get to the next step of attaching wheels, either of two experiences will be helpful—knowing something about the potter's wheel or about the rolling logs.

We have presented three accounts of the wheeled cart's origin: potter's wheel, rolling logs, and sledge. Which one, if any, is correct? No single answer can be given with confidence. However, something is clear and remarkable: all accounts employ the same ideas—complementarity and repetition. We have rediscovered two corresponding heuristics:

The Complement Heuristic Combine those inventions that have complementary evaluations. For example, the potter's wheel moves readily with a minimum of friction, but it will not carry anything to different places. The vehicle platform moves with difficulty because of friction, but it can carry large loads. Combine the two to achieve the capability of carrying large loads without much friction. Similar reasoning holds for the use of a rolling log or a runner.

The Repetition Heuristic When a part or principle works well, we should try to use several copies of it. One potter's wheel works to reduce friction, so try several. And similarly for the passage from one rolling log to several, from one skid to several.

What is most interesting here is that all accounts of the wheeled cart's origins have in common these two heuristics. Again both heuristics are vague, but when put together they begin to have real power. Yet the combining process is not so easy that it will yield completely to two such general rules. In all these ancestral paths to the wheeled cart, an explicit axle must be added, and a way must be found to fasten it and the wheel to the vehicle. These are momentous steps. They seem simple to us now, but for minds not biologically different from our own, this connection took tens of thousands of years.

We are now ready to state another important discovery principle:

The Comparative Heuristic Look at important inventions (like the wheeled cart), try to find rival accounts of their development (or make some guesses), and then extract the underlying heuristics behind those accounts.

Let us take stock of what we have done. We have used a comparative analysis to extract not next steps on an invention path but new heuristics. These heuristics come from abstraction and generalization based on important inventions, or transitions that took place within an invention family. To the extent that they are useful, they give us a new understanding and appreciation of historically important inventions. A suitable test of their usefulness is their power to help us comprehend, order, and appreciate what developed to change the course of an invention path.

But shouldn't we worry about promiscuously generating heuristics? Yes and No. Obviously, they should not be generated willy-nilly. But there is no need to worry as long as the heuristics have generality beyond the immediate setting of their discovery, and as long as their number is substantially less then the number of inventions they purport to account for.

The Eye of the Needle

I have a special fondness for needles because some of my earliest thoughts on invention date to thinking about the ideas embodied in sewing. Let me reconstruct that path.

I'm looking at a modern eyed needle (Figure 12.3). Actually, it is not so modern, dating back at least 25,000 years to the Cro-Magnon caves of Europe. Sewing is a momentous human invention. The eyed needle and its precursor inventions allowed for the construction of clothing and shelter, and the synthesis of complex artifacts out of more simple components, the standard activity of invention. A recent review of the Museum of Natural History's exhibit "Dark Caves: Bright Visions" asserts that sewing is one of the great unheralded inventions of human history. The Paleolithic bone, ivory, or antler needle, united with a sinew or hair thread, produced a new capability: the use of raw materials (such as hides, barks, grasses) to make novel composites. More than one of our Nobel Prizes for early invention is warranted here, and surely there are more heuristics to discover by unraveling the possible path of this invention.

The different implements in sewing's history suggest an interesting idea: The change from one form of sewing to another may be due to some advantage gained by the shift. Moreover, there may be interesting heuristics that underlie the advantage. Putting these ideas together introduces a new heuristic idea—*gain analysis,* the study of the gains (and sometimes the loses) as we follow an invention path. The trend or movement along an invention path enables us to analyze where we have been and where we may be going. Said in another way, the idea is to examine an invention at different points in its historical development to extract the underlying heuristics of change.

The modern needle

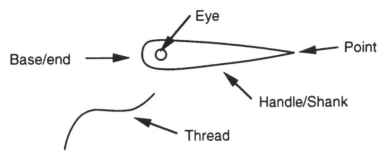

Eye

Base/end →

Point

Handle/Shank

Thread

Figure 12.3   Modern eyed needle and its parts.

Contemporary hand sewing employs a composite tool, the base-eye needle, which has multiple functions. The principal parts of the needle are shown in Figure 12.3. The breakdown or parse reveals a durable tool, the needle, and an expendable material, the thread. Each has important subcharacteristics.

The needle consists of a series of important parts and functions. First, there is a point for penetration, with the point as the working edge of the needle. Next there is an eye that acts as a carrier and feeder of thread. As such, the eye is the interface between the needle and the thread. The base of the needle serves as the place of contact for pushing the thumb against the needle; more generally, the base is one part of the human interface. The handle or shank is a place for gripping the needle; it is the other part of the human interface. A precision grip is required for manipulating the entire needle; the grip involves coordination between thumb and forefinger. In a tool as simple as the eyed needle, we note three forms of interface: two for the hand (the base for pushing and the shaft for pulling) and one for connecting needle and thread (the eye).

The thread also needs a series of properties: flexibility, strength, durability, reasonable uniformity, and low friction to allow for insertion in the eye and for travel through the material to be sewn. It must have reasonable length and be readily cut. Possible early candidate materials include hair, sinew, rawhide strips, and vegetable fiber.

The use of the needle and thread for sewing also involves a number of important function and action pairs. The function of threading the needle is achieved by pushing thread into the eye, pulling it through with a precision grip in which the actions of thumb and fingertips are coordinated. We penetrate the materials to be sewn by pushing the point through the "fabric," with a precision grip on the shank. Pulling through is accomplished by pulling on the shank with a precision grip to bring the needle through the material.

To discover the needle's precursors, I parsed it and its use into parts, functions, and actions, as above. Next, I examined the historical record for ancestral inventions with matching components or functions. The examination revealed an *awl*, a *fork*, a *joined awl-fork*, a *middle-eye needle*, and a *notched needle*. Each of these is shown in Figure 12.4. All these forms have been discovered at various archaeological sites, but not all at one site. The awl has the penetration function of the needle, the same kind of handle/shank and base, and some of the same kinds of required action. What it lacks is the thread-carrying function provided by the eye. A fork, used in conjunction with an awl, allows thread to be pushed through a perforation made with thean awl.

The *joined* awl-fork is essentially the integration of its precursor components. By now we know that joining together two or more tools is a common path of development in invention. The composite awl-fork allows the same procedures and sewing functions as those available with its separate components, but all in the same tool. Having both the awl and fork functions joined in the same tool eliminates some of the switching processes involved in changing between tools as they are successively picked up and put down.

The *middle-eye needle* may be regarded as developing from the pure fork, with the thread-carrying capability migrating up the shaft. The middle-eye needle is an improvement over the separate awl and fork, because it is an integrated tool that, as a bonus, provides a thread-carrying interface. Despite its advantages over earlier forms, the middle-eye needle also has several negative aspects: The point of maximum stress is at the middle of the shaft, the weakest point because of the placement of the eye. Also thread coming out of the eye is likely to

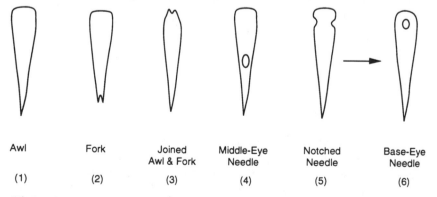

Awl	Fork	Joined Awl & Fork	Middle-Eye Needle	Notched Needle	Base-Eye Needle
(1)	(2)	(3)	(4)	(5)	(6)

Figure 12.4   Precursor candidates for the eyed needle, (1) through (5), and the eyed needle (6).

bind the passage of the needle as material is sewn. Both of these problems are corrected by moving the eye up the shaft toward the base of the needle. The standard base-eye needle is the likely result.

The movement of a feature, such as an eye from the middle to the base end, undoubtedly provided better functioning. The rearrangement of features for better functioning is a fine-tuning, a common feature of invention in which the arrangement and layout of realized parts is shifted and shaped for optimization.

The *notched-end needle* affords still another possible path to the base-eye needle. The notch is an alternative interface between needle and thread. Its advantage is in having the thread interface at the base where it is relatively out of the way. Its disadvantages include possible friction or binding as the needle is pulled through, and possible breakage produced by friction forces on the weakened notch end. However, the use is simple: tie the thread in the notch and then sew, much as you would with a base-eye needle. Notice the easy presumption of continuous sewing for the notched needle; it is hard to imagine using it for making discrete stitches, where each stitch is cut and tied separately.

Of the various sewing implements presented in Figure 12.4, the fork is the one for which the least evidence is available. In fact, archaeologist Danielle Stordeur-Yedid interprets the fork as a broken needle, but that is not consistent with the shape she depicts. Nor is it consistent with developing the joined awl-fork, an implement discussed by another archaeologist, François Bordes. Indeed, the joined awl-fork is complex enough that it is likely to have its own precursors, like the separate awl and fork, of which it is the integration. On these grounds, the fork seems likely to have been a real and important sewing implement.

Using the principles we have developed so far, let us construct some plausible invention paths to the base-eye needle in its modern form. One plausible progression, using the numbers of Figure 12.4, is: (1) + (2)→(3); the awl and the fork are combined to produce the joined awl and fork. That is what we would expect, based on our prior discussions of joins. The next alternatives are less obvious and include (3)→(6), and (3)→(5)→(6). Both of these paths involve *principle hopping*, finding a new way of getting the thread carried by a pointed object. The first starts with the joined awl and fork and moves directly to the base-eye needle, a very large step indeed. Not only does the form of the needle change but the underlying form of sewing may also have changed. The second progression, (3)→(5)→(6), seems more plausible. Moving from the joined awl and fork to the notched needle looks like a smaller step. The notched needle must imply sewing with a continuous thread, so the underlying mode of behavior need not change in going to the base-eye needle. However, the move from the notched needle to the base-eye needle is still a substantial step. The spoiler in all of this is (4) the middle-eye needle. Where did it come from? A possibility is: $(1) + (2) \rightarrow (4)$. In this account, the middle-eye needle is a solution to the problem of alternating between awl and fork. If so, it was a big step. Then from (4) to (6) is a matter or refinement. As the hole migrates up the shaft, the needle becomes stronger and the thread produces less friction. Perhaps, after a number of generations of nondeliberate variation, the eye ends at the base of the shaft.

At first glance, these forms of sewing may all be classified as simple evolutionary drift where deliberate invention does not really play a role. According to this view, individual chance variations in the production of an implement are selected for efficiency, and through that selection an optimization of function is finally reached. The migration of the eye up the shaft may be accounted for this way. Those eyes that are higher on the shaft work better (break less often and offer less friction) than those that are lower. After several generations of nondeliberate variation, the eye ends at the base end of the shaft.

This is an evolutionary drift argument applied to invention, one based on random variation and selection. Yet it fails to explain the development of an invention as simple as the needle. No evolutionary account based on natural variation and selection *along a dimension,* and the corresponding selection, can produce a new feature, such as the eye. The eye is a topological discontinuity from its precursors, which did not have interior openings. Nor could an evolutionary drift account produce the joined awl and fork, because there is no single underlying natural dimension of variation. The join or combination is new. Design space is not continuous, particularly in the joining of discrete forms. However, implicit evaluation criteria may well be continuous: speed and flexibility of sewing, strength and resistance to breakage and repair, and the time spent on switching processes. The invention pathways from precursors to the base-eye needle, in Figure 12.4, were movements away from discrete processes and movements toward continuous processes, but the design changes themselves were often discrete. Therefore, an evolutionary argument, based on manufacturing variation and selection, falls short. What we have here is likely to be a set of alternative paths to the same invention, the modern form of the eyed needle—some forms produced by selecting from unintentional variations in manufacture, but more by acts of invention that go beyond selecting from the chaff of chance variation.

So far, the discussion has centered on the tangible tools of sewing. Yet that is only part of the story because different sewing *procedures* may be associated with the different artifacts. Those procedures are as much acts of invention as the implements they are associated with. Let's look at some of the procedures in more detail.

The first and most elementary procedure is sewing with an awl and fork in a *discrete processing mode*. It involves the use of an awl to poke holes in material and then a fork to push through a short piece of thread that is tied separately for each stitch. This sequence of procedures constitutes a discrete process in which a series of distinct steps is completed for each stitch before the next stitch is begun. Each stitch requires the following steps, but not necessarily in the order given. Pick up a cutting implement, like a knife or scissors, and cut a short piece of thread or fiber; then put down the cutting implement. Pick up the awl and poke a rather large hole through the two edges of the material to be joined; lay down the awl. Pick up the thread and place it in position over the hole just made, holding it in place with the thumb. Pick up the fork and use it to push the thread through the hole; put down the fork; and tie the thread. Then repeat or loop through the procedure for each successive stitch. In its most primitive form, the whole procedure requires the looping or repetition of a discrete process (one tie or stitch at a time), until the two edges of material are firmly attached in a scam.

While this particular form of sewing seems awkward, some modern forms of sewing approximate it. The surgeon who is suturing a wound is using a discrete stitch form, but with a curved base-eye needle. She takes a single stitch at a time and ties it before going on to the next stitch. The discrete mode allows for attaining maximum control of tension for each individual stitch. But it is an effortful, time-consuming mode that is unnecessary for the fabrication of clothing.

With some of my students, we clocked the time it took to sew in this mode. We first made our own awl and fork of bone, and unlike the sinew or hair of old, we simplified by using regular thread. Table 12.1 shows the time per stitch in sewing together two pieces of cloth.

Table 12.1   Sewing Time in Seconds per Stitch (2 to 4): Experienced Sewer[a]

Component	Discrete Awl & Fork	Continuous Awl & Fork	Continuous Base-Eye Needle
Total	38.3	26.4	5.3
Processes[b]	28.9	16.6	5.3
Switches[c]	9.5	9.8	—

$N = 1$.

[a] Similar differences also occurred for a naive sewer, so it is unlikely that experience in sewing is an important factor here.

[b] Direct processes involve operations connected with constructing a stitch, such as penetration or pulling or pushing thread through; contrast with switching processes mentioned in note (c).

[c] Switching processes involve putting things down, looking for them, picking them up, in short, dead time between operations.

The total times are separated into processing times and switching times. The first is concerned with actually making stitches and the second with preparation times like putting down one tool and picking up another. The results show that processing time for this mode of sewing is indeed lengthy, 28.9 seconds per stitch. But so is switching time; all that picking and putting down and looking for the right tool took about 9.5 seconds per stitch.

A reasonable guess is that tools and procedures that are more efficient, that reduce switching between movements and actions, will produce greater efficiency. To this end, another sewing procedure called *batch processing* was tried. In using the awl and fork for batch processing, all operations of one kind are completed before going on to the next operation. All the fibers or threads might be cut to length, with the knife or scissors picked up and put down only once, at the beginning and ending of the cutting. Then the awl's penetration function was applied repeatedly: instead of punching one hole at a time, all were punched before doing anything else. Again, the awl need only be picked up and put down once for an entire seam. Batch processing reduces effort by eliminating unnecessary switching between tools, actions, and underlying mental routines.

Early on, we found a problem in applying batch processing to sewing. It was impossible to keep everything lined up properly for a long seam because the holes we poked sealed up before they could be threaded and individually tied. For this reason, we were not able to realize a pure batch processing mode of sewing. But batch processing is a very common form of manufacture.

Next we tried the awl and fork with continuous processing. In a *continuous processing* mode there are no sharp demarcations between steps; one leads into another. Instead of cutting the thread into short sections, we tied it at one end of the seam, poked holes, and then pushed the other end successively through the fabric until the seam was finished. The time per stitch in this mode is also shown in Table 12.1. Because the same tools are used, the switching processes are essentially the same, and there is no time reduction here. But the continuous thread process is much more efficient than the discrete process, so the time drops sharply to little more than 16 seconds per stitch.

A trend is emerging. The continuous thread mode is leading us away from a series of discrete or batch processes to a manufacturing process with fewer steps and briefer alternations between steps. However, remnants of discrete processing remain. There is a separate tool (the awl) for punching holes and another tool (the fork) for pushing the thread through the already made holes; and these must be alternately applied. What would happen if we integrated more tightly some of the separate inventions?

Accordingly, another mode of sewing to consider is one that uses the awl and fork joined into one tool. The sewing processes here parallel the discrete and continuous processes of the separate awl and fork, but now fewer picking up and putting down operations are required. We did not directly test this mode, but it certainly looks promising.

Another historical mode that we did not test is sewing using the middle-eyed needle of Figure 12.4. The penetration or punching function of the awl is still present, and in addition, the function of the fork has been combined with the awl and improved upon. The eye is effectively an interface between the awl and the thread; it acts as a carrier of the thread. While one could use the middle-eye needle for discrete stitches, it also allows for fully continuous processing, thereby reducing switching time.

In the modern base-eye needle, discrete and batch processes are readily replaced by continuous sewing in the full sense. Penetration and thread-carrying are now integrated into the same tool. Many steps have been eliminated. Also the seam sewed is tighter, because as sewing becomes more continuous and faster, it is easier to place stitches closer to one another. We tested continuous sewing with the base-eye needle, and as Table 12.1 shows it is much faster than either of the other modes. Switching time from picking up, putting down, and looking for tools is completely eliminated, and the integrated processes of penetration and thread carrying greatly speed up sewing, with about 5 seconds per stitch. The base-eye needle is a masterful combination, certainly worthy of a Nobel Prize for ancient technology.

From the needle's path of invention and the modes of sewing described here, the following heuristics may be abstracted:

- Determine what the underlying evaluative criteria are that drive the invention process. (Rapid assembly, with minimal effort.)
- Increase efficiency by moving from discrete to batch to continuous processes. (This is by far the earliest example that I know of a continuous manufacturing process.)
- Minimize breakage and maintenance processes. (Move the eye up the shaft.)
- Integrate different tools (awl and fork) and processes (penetration and thread carrying).

These are specific heuristics discovered by looking for the driving principles underlying sewing's invention path. Two general ideas are basic to these heuristics. One is the gain from fewer switching processes and less switching time. The other is the efficiency that can result as we move from discrete to batch to continuous processes. Evaluation criteria like minimizing or eliminating the time it takes for different processes tell us in what direction to move as we navigate in an essentially unlimited space of possible next steps. Notice that even though these heuristics were arrived at historically, across many artifacts and many minds, they may be incorporated into the mental makeup of the individual inventor or designer. But are not these principles "obvious"? To charge that they are is to overlook the fact that it took thousands of years to move from the continuous processes associated with the base-eye needle to similar continuous processes associated with the assembly line, as we will see a bit later.

We can learn a final lesson from the invention of the eyed needle and the wheel. These were big inventions, so what distinguishes present-day invention from that of our distant forebears is not a matter of intelligence. I know of no evidence that our Cro-Magnon ancestors were mentally less facile than we. Certainly, the inventors of the wheel were our equals in mental hardware. Nor does it convince me that we invent by deliberate action while our ancestors did it by random variation and selection. I believe the difference is that we have more invention models and principles to draw on, more heuristics to lubricate our mental paths than did our ancestors. And that can make all the difference.

13

Applying Heuristics: Inventions After Their Time?

Some time ago, a friend and I were having lunch. As was our custom, we tossed around several ideas to speculate about. The one we settled on was the origin of the phonograph. We knew that it was a radical invention, but how much so? Was it an invention before its time, an invention on time, or an invention after its time?

We are all familiar with the idea of inventions *before their time*. These are the inventions of science fiction and speculation: The wrist radio of an earlier day, the space ships of the Star Trek sagas. A splendid historical example, discussed earlier, is Leonardo's helicopter. It is a brilliant concept, but it could not have succeeded in his day. To be practical, the helicopter needed a small, light, and powerful engine, and that was centuries away. In addition, key aeronautical ideas of lift and control were needed, and they too were centuries removed from Leonardo's time. It wasn't as though people didn't want to fly or had no purpose in doing so; to fly is an age-old dream that goes back at least to Daedaleus and Icarus. If Leonardo's helicopter had worked, people would have known uses for it. That was certainly the case with the Montgolfier brothers and the first manned hot-air balloon ascent in 1783. Soon after the balloon's launching, uses for it were found, from scientific and military observation to entertainment.

People are not surprised that some inventions, like the helicopter, are before their time. Certainly, they should not be surprised by an *invention on time*. But what does that mean? It means that an invention appears shortly after the essential components, principles, and mental models for it all become jointly available for the first time. And a need or desire for it is also present—or can be readily instigated. Everything comes together and, shortly after, the invention appears.

An example is in order. Suppose that a transition from stone knife to bronze knife has just been made. The new techniques for bronze metallurgy are now in place, and people know how to form alloys, make molds, and cast knives. Then using the same techniques to cast another tool such as an ax head—which already exists in stone form— is an example of an invention on time. Such an invention is a natural outgrowth of what has gone before, including knowledge, components, and procedure. Moreover, the metal knife already exists as a mental model of the new invention class.

If the idea of inventions before their time or on time makes intuitive sense, people are often surprised by the third case to consider, *inventions after their time.* In fact, I want to argue that some important inventions took longer to develop than they should have: they were inventions after their time. It is important to find out why. More than a thought experiment, such inventions could provide important lessons. Let's look at some cases.

The Phonograph

Thomas Edison's phonograph of 1877 is usually regarded as a totally unheralded invention. Certainly it is an invention before its time. Or is it? In Edison's day, no open-ended market surveys of consumer needs would have listed a device with the capabilities of the phonograph. Storing and reproducing sound had not been a problem that people were eager to solve. Indeed, it had not been a problem at all; no one *needed* a phonograph. So Edison's invention at once created a problem and solved it. That is what *unheralded* means in its most radical sense.

But why was the phonograph unheralded? Should it have been? Indeed, could it have been invented earlier? To be provocative, let us say Yes—it was an invention *after its time.* Now let's see why.

We can easily find several explanations for the phonograph's late coming. The first to consider is *the absence of critical components.* Perhaps some critical component was missing until shortly before its development? Such an explanation is certainly appropriate for the timing of other inventions. As we have seen, controlled, sustained, and powered human flight did not occur before 1900, because success required a small, light, and powerful engine and greater knowledge of aeronautical principles—things not available until the time of the Wright brothers' efforts. Perhaps the phonograph also could have been invented earlier if it were not for some missing ingredients?

Let's work into answering this question by first considering the structure of the phonograph as Edison invented it, with the original idea occurring in 1877 and the patent issued in 1878. Figure 13.1 is a sketch of Edison's phonograph. The components consist of a long helically threaded cylinder (grooves) around which thin metal foil is wrapped, a horn-like cone, with a stylus attached to a diaphragm in the cone, and a crank. Sometimes a flywheel was also included at the end opposite the crank to provide smooth rotary motion. This version has separate "microphone" and "speaker" structures. An earlier version used a single common structure for both functions.

For recording, one yells into the microphone and turns the crank. The cylinder rotates and the stylus makes contact with the foil. As sound waves move the diaphragm in and out, the stylus makes corresponding indentations in the foil. The sound wave is encoded as a time-locked impression of varying depths in the helical track around the metal foil on the cylinder. For playback, one rewinds the cylinder with the crank and places the stylus of the speaker in contact with the beginning of the groove in the foil. When the crank is turned, out comes the recorded sound. One of the early tests was to record and playback "Mary had a little lamb. . . ." The early versions were capable of recording about ten seconds of sound.

Now to return to the question of whether the phonograph was so long in coming because the critical components were not developed earlier. If a parse of the phonograph's parts is made, together with an approximate date of first appearance, we come up with the parts and dates indicated in Table 13.1.

Noteworthy here is the mix of ingredients. Who would think ahead of time that drum head diaphragms and metal foil would enjoy commerce with screws? Nonetheless the principle point is clear: All the parts of the phonograph were readily available and had been so for a very long time. Unavailability of components cannot explain the delay in the development of the phonograph. On the sole basis of component availability, the phonograph should have been developed before its actual invention date of 1877. However, all we have shown here is component availability. Inventing by haphazardly assembling components is very unlikely; the number of possible component combinations assembled at random is astronomical. So even more important to the development of an invention such as the phonograph may be the underlying concept or idea on which it rests.

Hence a second explanation to consider for the phonograph's late development is *the absence of a mental model or concept.* For example, nothing in the way of an existing mental model may have existed to suggest the possibility of recording, storing, and playing back sound, and therefore the possibility of the phonograph. If there were no mental models, then it is not surprising that the phonograph didn't surface earlier. In fact, there were at least two existing models for stored sound,

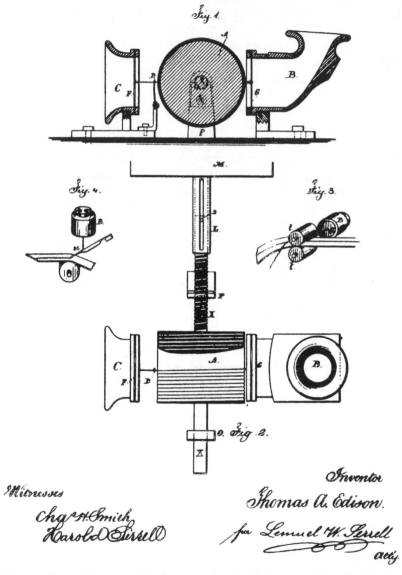

Figure 13.1 Edison's early phonograph. A cylinder covered with metal foil, and with a separate microphone and speaker. (U.S. Patent 200,521 of 1878.)

Table 13.1 **Phonograph Parts and Time of First Occurrence**

Part	Time of First Occurrence
Helical thread	Screw, Archimedes or earlier; 3rd century B.C.
Metal foil	Ancient
Horn	Centuries old
Vibrating diaphragm	Drum head; ancient
Continuous rotary motion	Potter's wheel, 3500 B.C., or before
Fly wheel	Potter's wheel, 3500 B.C., or before
Crank	About A.D. 800

as well as a related analogical model based on light. The two models for sound are biological memory and the music box. Biological memory immediately suggests all the functions of the phonograph: recording (in a conversation you can remember what you have just heard); storage (until you are asked to remember the conversation, the information must have been in storage); and playback (now you are recalling the conversation). The other model for stored sound is the music box. While the mechanism of the music box is not promising as a way of leading to the phonograph, it does indicate the possibility of representing a sound code in another medium.

An analogical model for stored sound based on light was also available. An existing invention, photography, was in place fifty years before for the medium of light. Notice the closeness of the analogy between sound and light: encoding/recording (expose the image to a light-sensitive surface or plate); storage (develop and fix the image); playback (print the negative). If one were to generalize from the medium of light to the medium of sound, then all of the ideas critical to the development of the phonograph are analogically suggested.

The conclusion is inescapable. The delay in the development of the phonograph cannot be explained by an absence of either mental models or record-store-playback concepts.

A third explanation for the delay of the phonograph is *the lack of invention heuristics that are systematically applied.* I believe this explanation best accounts for the delay in developing the phonograph. And it also explains the occasional failure of even the best minds in seeing deep opportunities for invention.

Before further explanation, you might want to speculate on how the phonograph was actually developed, how the idea for it came about. Try to use the invention principles we have developed.

This is the place that my friend and I were at over lunch. Our goal was to provide a plausible account of the phonograph's origins. At the time, we were ignorant of the actual origin, except for Edison being the inventor and the approximate date. We did not know anything about the original configuration in Figure 13.1. Yet reflection revealed the big ideas involved: recording, storage, and playback of sound. We then probed our limited memories for the history of inventions related to sound, the underlying medium. Those memories indicated that for some time before Edison the technology was present for recording sound waves on a revolving drum or kymograph that was covered with lamp blackened paper.

For example, Leo Scott's phonautograph of 1857 used a smoked drum kymograph setup similar to that in Figure 13.2 to produce tracings of sound waves. (Koenig's 1859 version is shown in the figure.) The kymograph drum was rotated rapidly, and sound waves entered a cone or horn with a diaphragm in it; a stylus was attached to the diaphragm and through a lever system made contact with the smoked drum which was rotating on an internal helical thread. A tracing of the sound wave was etched on the soot-covered paper. Such tracings are *visual representations* of the wave form produced by the sound. Moreover, the device itself had a structure and appearance similar to Edison's phonograph that would take another twenty years to develop. Some evidence indicates that Edison was familiar with this device, but not at the beginning of his sound studies.

Figure 13.2 The phonautograph, an early device for recording sound. (Drawing by M. Sheldon; based on Miller, 1934. The exact nature of the stylus attachment is unclear.)

What is missing from the phonautograph? Essentially, playback capability. At our lunch, we had a general idea of how the phonautograph worked, if not the details or the exact time frame. Our central speculative insight was that playback could be considered as the *inverse* of recording the visible trace of the sound wave.

Application of an inverse heuristic immediately suggests the use of the kymograph's visually recorded sound wave to drive, in reverse, the processes that produced them. Going backwards and starting with the visual trace, the wave form somehow drives a stylus connected to a lever system, which then moves the diaphragm in the horn-microphone. The result is a microphone turned into a speaker that emanates sound waves. Still there is a problem. How can we have a stylus track a *visual* wave form that is recorded on a drum? One of us then remembered that early phonographs used *waxed* cylinders. That suggested that the waves had been recorded not on the surface or lateral plane of the cylinder but in the depth plane—that is, by indentations of the stylus in soft wax, resulting in a groove of varying depth. Then playback could be accomplished by having the stylus driven up and down by the groove's indentations in the wax. We had difficulty with an explanation for the cylinder driving mechanism (the hand-turned crank), because we were trying to think of something more sophisticated. Nonetheless, we left lunch with a warm glow, convinced that we had constructed a plausible account of how the phonograph came to pass.

Later I looked up the historical record. Our physical account of the way the phonograph worked was essentially correct, with a few exceptions. We had not anticipated the use of either the foil or the crank. Our psychological account was entirely wrong, however. At least initially, Edison's idea did not in fact come from thinking about an inverse. Instead, its origin was inspired by serendipity and analogy.

Edison had been working on improvements in the telegraph. He wanted a repeating telegraph, one that could store messages and then at a later time send them to several stations at once, to produce maximum use of a telegraph line. Perhaps there was a queue during the day, and full capacity was not used at night. With Edison's repeating telegraph, one could record the telegraph messages off-line during peak hours and then send the recordings during the slack time at night. An early version of the repeating telegraph consisted of a disk on which a series of dots and dashes were embossed on a spiral track to represent a message. So we have here a recording and storage system. For playback, all that was needed was to rotate the disk under some contact points that on striking bumps activated a telegraph circuit and sent the recorded message to as many stations as wished.

According to one story, Edison inadvertently played the disk back at a faster rate than usual. As a result he heard something like sound emanating, and that suggested to him the idea of recorded sound. This then is invention by serendipity and analogy: He was lucky to hear the sound, but once heard he could draw on the analogical ideas of record, store, and playback from the telegraph as he applied them to sound. Another story has him working with the transmitting diaphragm of a phone receiver. Because he was partially deaf, he detected sound by feeling vibrations. Once having felt the moving diaphragm, Edison thought of mounting a stylus on it to make impressions on a storage medium. Then during playback, the process was run in reverse, with the impressions activating the stylus and ultimately producing sound waves through the actions of the moving diaphragm. Perhaps this thinking occurred in the context of working on the repeating telegraph, again providing an account based on serendipity and analogy.

In any case, he had the idea of recording and playing back sound. He attempted to do this in several ways. One was with a wax-coated paper tape moving under a stylus. The sound waves then left a negative impression on the tape. For playback the tape was rewound and run back. The peaks and valleys now drove the diaphragm up and down, sending out a faint sound signal. Unfortunately, the waxed paper was good for only about one pass before it lost its embossing.

At this point, Edison may have looked to a device like the phonautograph and correctly read the possibility of applying an inverse to the process. Certainly, his phonograph of Figure 13.1 bears a striking resemblance to the phonautograph of twenty years earlier. Alternatively, he may have moved directly to this system from the paper tape, because he was very fond of cylinders.

Which account of the phonograph's evolution is more direct and serves best as a guide to future invention—one of the historical accounts based on serendipity and analogy or the account based on finding an inverse for the phonautograph? To me, the inverse account seems more natural, more direct, and more generalizable. If we accept the inverse as a plausible invention path, let us codify our hard-won knowledge and hindsight with by now an old friend:

> *The Inverse Heuristic*   We should be on the lookout for interesting phenomena that have missing inverses. At least some of those missing inverses may be worthwhile. Being able to play back the representation of a recorded sound is clearly important, once you think of it.

I am not suggesting that the phonograph could have been invented *merely* by applying an inverse heuristic to an old device for recording sound. Very considerable skill was required to put it all together. The early phonographs did not work well at all. A long period of imaginative fine-tuning was required to bring the mechanical phonograph to practical implementation. Nor is the application of inverses likely to be helpful to one unfamiliar with a related problem or having access to the materials needed for a solution. Nonetheless, in the right hands, looking for missing inverses is a useful guide to invention.

Why is it that some earlier investigator of sound did not think of simply applying an inverse heuristic to the existing technology of recording sound waves? Certainly, Scott or Koenig had a good shot at the phonograph twenty years earlier. Even if that was not their purpose, at least a mention of the idea would have been in order.

The answer for the phonograph's delay that makes best sense is a simple one: Invention heuristics are not part of the codified knowledge of scientific literature and textbooks, and each individual is left to discover anew important heuristics, such as looking for missing inverses. As a result, heuristics are not readily or systematically applied when inventing, which is unfortunate, since they can be quite useful in generating ideas.

Before drawing lessons, a comment on hindsight is in order. I am not using hindsight to suggest that Edison or the early investigators of sound were mentally slow—far from it! Instead, hindsight analyses allow us to pinpoint where, in retrospect, even the best thinking was unsystematic or had gaps. Hindsight tells us how we can build on and regularize our thinking by making explicit the underlying heuristics of an invention path.

I believe the main lesson to draw here is that the phonograph is an invention after its time. The reason it occurred later than it needed to is not because of missing components or a missing mental model. It is because simple heuristics were either not known or were not systematically applied.

Velcro

If the thinking that guided us in claiming the phonograph was an invention after its time is to have persuasive force, we must show it has generality across other inventions as well. Perhaps, after all, the phonograph is an invention with unusual characteristics that are not generalizable. Perhaps there is no broader category of inventions after their time?

To test the generality of this claims, consider another example, an invention that also begins with serendipity and analogy as the basis of discovery. We will then examine a more direct path to the same invention. I want to argue that the path not taken for Velcro could have been; and, if it had, Velcro would have presented itself to the world much sooner.

Before going further, you may wish to draw a picture of what Velcro looks like under a magnifying glass. Suppose that you were walking in the woods and came back with a sweater covered with burrs. How is it that the burr fastens to the sweater? This is the problem the Swiss inventor George de Mestral was confronted with after he returned from a walk in the woods, covered with burrs. When he looked in his microscope, he saw a miniature hook-eye system: The burrs had the hooks and the wool of his sweater the eyes. What we have here is an origin account for Velcro based on a found-in-nature analogy. Although the insight for using the burr principle as a fastener came rapidly, it took de Mestral nine years to achieve a viable product. He spent most of the time searching for suitable materials and developing manufacturing equipment and processes. That nine years we will not dwell on. Our question is a simple one, why did we have to wait until the 1950s to reach the idea of Velcro?

Again, consider some rival explanations for the late occurrence of Velcro. Perhaps other people had also noticed the same phenomenon, burrs sticking to something. Perhaps they had even thought of a fastener based on the burr, but they rejected the idea because there was *no suitable material.* In this view, the invention of Velcro had to await the invention of nylon by Carrothers in the late 1930s. Against this argument, we may point out that Velcro is also made of other materials, including steel hooks and eyes; steel Velcro is used in high temperatures or other hostile environments. The point is that steel has been around for a long time. Evidently, the delay in the Velcro idea cannot be ascribed to the absence of a particular material. So why didn't we have Velcro much earlier, made of steel or some other material?

What other way could Velcro have been invented? It is time now to test your intuition of how Velcro looks under a microscope. A schematic version is shown in Figure 13.3. Examination of the figure reveals a miniature hook-and-eye system. The surface with hooks is pressed into the surface with eyes; because both are flexible, they become entangled.

Now individual hooks and eyes are an old idea. They have been used for centuries as clothing and boot fasteners. The presence of hooks and eyes as the basis of Velcro indicates that artifact as well as biological models for Velcro were not lacking.

What is new about Velcro in comparison to the established hook-and-eye clothing or boot fastener? Several things. The size scale of Velcro's hooks and eyes is very small, so small that many users of Velcro don't even know about its underlying hook-eye structure. Why is its size so important? Because a small size requires imprecise or indeterminate matches between hooks and eyes. This is a radical idea. Traditional uses of the hook-eye fastener—for example, those on boots—had each hook matching with a specific eye.

To have hook-eye matches at the micro-size level, it is necessary to expand the target area. Traditional hook-eye fasteners are each in

Velcro

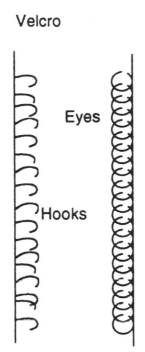

Eyes

Hooks

Figure 13.3 A schematic representation of Velcro. Hooks and eyes at a miniature scale, with the additional requirement that each element be flexible.

one-dimensional linear arrays, and then the two arrays move toward one another to match in lockstep, one to one. When the scale is very small, the match of individual pairs of hooks and eyes is not easy and their embrace is not strong. The solution is two-dimensional arrays, one with hooks and the other with eyes. Then two *areas* move toward one another as much easier targets for alignment—and much stronger too.

The collected differences between hook-eye systems and Velcro are listed in Table 13.2. First, there is a change in scale. As noted earlier, this is a common invention ploy. Because the scale changes, we must extend the hook-eye system into two dimensions. Finally, in Velcro, the match of an individual hook with an individual eye must move from determinate to indeterminate. At least some hooks will mesh with some eyes, but it no longer matters which hooks go with which eyes.

From the three differences of Table 13.2 we may extract and codify a set of heuristics for transforming hooks and eyes into Velcro.

Table 13.2   **Relationship of Traditional Hooks and Eyes to Velcro**

Characteristic	Hooks/Eyes	Velcro
Scale	Moderate	Micro
Dimensionality	1-D	2-D
Matchup	Determinate (1-to-1 fixed correspondence)	Indeterminate (any-to-any correspondence)

The Scale Heuristic   Try changing the size of one or more components in an existing invention. Often a change in size scale will open up entirely new applications.

The Dimensionality Heuristic   Try changing the dimensionality of one or more components in an existing invention. A royal road to generalization in mathematics is to vary the dimensionality of a concept.

The Matching Heuristic   Try changing the basis of matching or linking between components, from deterministic to haphazard. The way components are linked together is a fundamental concept.

Together these heuristics assist the observant inventor in transforming traditional hooks and eyes into a product much like Velcro. They are simple principles for generalizing on a process or artifact. In fact, once the simple heuristic of changing scale is introduced, changes in dimensionality and matching are required to line up hooks and eyes. So the biggest part of the step is simply to think about hooks and eyes at a greatly reduced scale—a very general heuristic—and the others follow, even if the practical problems like mode of manufacture remain formidable.

Once more, we have an invention after its time, because the relevant heuristics were not applied and we had to wait for biological analogies—found-in-nature insights such as de Mestral's. But that sounds awfully glib, and it must be emphasized that the blind application of heuristics by someone unfamiliar with the materials and skills in an area is not enough. My claim is for the importance of heuristics, not their sole importance.

Qualifications and Caveats

But before arriving at a firm conclusion on the role of heuristics in invention, we should consider arguments against the approach taken here. First, a critic might charge that we have made the whole process of invention far too simple. It is all *after the fact;* anyone can make an invention path appear simple in retrospect. This critic would say our reasoning suggests that the early mechanical clocks of Europe and China were simply the result of stirring together the simpler machines of classical antiquity: the inclined plane, the wedge, the screw, the lever, the wheel, and the pulley. Of course, that is true—and ridiculous at the same time. The order of complexity required in going from the simple machines to the mechanical clock is immense. Many steps are anything but obvious. No small set of heuristics will explain so many intricate steps.

As a first defense, I am certainly not making so wide a claim. I only claim that in some cases the connection between existing inventions, the heuristics applied to them, and the resulting important new inventions are transparent.

As a next defense, I ask the critic if anyone wants to argue that there is no way of systematizing and speeding up the invention process. Not many people will wish to make that argument. So at worst we have a disagreement about the power of a particular approach for improving the efficiency of invention. That's not bad for a start.

Third, our critic may charge that we manufacture heuristics on the spot and may well ask "How many heuristics are there anyway?" Certainly, I have no answer. But we need not worry as long as the number of heuristics is relatively small in comparison to the number of ideas they help generate.

Fourth, our critic may continue, there is another way in which we may have made the process of invention too simple: The presence of the right nut and bolt in the right place can make the difference whether an airplane gets off the ground or not; if it does get off the ground, another nut and bolt may determine whether the flight is successful or crashes. So why paint with such a broad brush and make strong claims about the importance of heuristics, when we know that the process of invention is critically dependent on small-scale ideas, like nuts and bolts and specific knowledge and skills for assembling them in the right order?

There is merit in this critique, but it is not the whole story. If one looks at the sketches of great ideas in progress, like Leonardo's helicopter and the Wright brothers' drawings, many are conspicuously missing the nut-and-bolt level. That is the sub-subject matter level, the knowledge of materials and construction, that is often handled intuitively. It is important knowledge, but it is common to skilled practitioners and cannot afford the generality that a coherent perspective of invention requires.

Fifth, our critic may argue the heuristics used here are quite vague, from extremely general and low-powered ones like "change the size of a component" to at best middle-range ones like "find an inverse." That being the case, the heuristics are not of themselves sufficient to generate any ideas. There is no computer program that will find them very useful. That is probably true, and it returns us to the power-generality tradeoffs that we discussed earlier. Remember there are good reasons for believing that humans can do much better with vague heuristics than present-day machines can.

The more intelligent and knowledgeable the inventing system, the more capable it is of profiting from vague, general, and low-powered heuristics. Moreover, when heuristics are applied severally, the combination becomes synergistic. Looking for a needle in a haystack is minimally facilitated by poking in it with a small weak magnet; nor is shaking the stack to have the more dense needle sink to the bottom likely to help much. But when the two heuristic procedures are applied jointly, our odds of finding the needle can be expected to increase. They will increase even more, if other heuristic operations are also applied. For example, we might throw the stack on a swimming pool so the straw will float and we can drag the bottom with our magnet. Or absent a convenient swimming pool, we can burn the stack and drag the magnet through the ashes. Clearly, individual heuristics can be weak while in combination they are strong.

Sixth, a given invention can have multiple paths, as we have shown by the different possible origins for the phonograph and Velcro. Each of those paths then may have its own heuristics associated with it. Some people worry that a heuristically unique path to an invention does not exist, but I see that as no problem. The world of heuristic thinking is rich and messy.

Finally, to say that invention is propelled by heuristics is in no way to say it is a mechanical process. All the wonders of human memory and intelligence are required to extract heuristics, to know when to apply them, and to fill in the gaps and specific knowledge that is often required. By definition, the use of heuristics will not lead with certainty to new and important inventive ideas. At best, heuristics help us in knowing where and how to look. Often that is enough when an ordered memory and a prepared intelligence are at work. A good way to order memory and prepare intelligence is to study important historical cases of invention, like the phonograph and Velcro.

One of our most impressively human characteristics is the power to learn from experience. And one of the best ways of learning from experience is through the codification and systematization of hindsight. If we could not learn about invention through hindsight, we should be surprised. Indeed, recent technological history indicates an immense increase in the rate of invention. This accelerated pace may result from becoming more sophisticated about the *idea of invention—* its general implementation through a variety of principles and heuristics.

Historical case studies like the phonograph and Velcro allow us to recapture candidate processes of invention. Those cases may then be played back again and again, to better understand the underlying principles of creativity and invention that are involved. Notice that we are now using Edison's phonograph as a metaphor for the hindsight method: recording and playing back the history of great inventions to discover underlying heuristic principles to guide the creative mind. To further develop the metaphor: Our "sound source" (the idea or case study) comes with much noise (misleading, faltering steps, omissions or skips in the creative processes, not to mention stuck grooves). What the hindsight analysis does is to filter some of that noise to extract the important "signals" (principles of invention) coming from these momentous creative acts. By listening to those signals and heuristics, perhaps we can amplify and systematize the creative signal, minimizing the lag for inventions after their time.

Let's take stock of where we are. In the last two Parts we have developed the ideas of describing inventions and applying heuristics to generate new ideas. It's now time to look at some important categories of invention, ranging from interfaces to transgenic forms, that illustrate these ideas. While the analysis of these inventions draws on the concepts of frame description and heuristics, I don't want to hit you over the head with those ideas. Hence, the rest of the book is self-contained and will not explicitly mention frames and heuristics; but if you wish to, the translation is easily made and you will see how my line of thought moved regularly between description and heuristic.

IV

COMMON
INVENTION THEMES

14

A Material
World

I can remember as a child experiencing a great thrill when a family friend gave me an arrowhead. Its flint texture and rough shape held the secrets of history in a way that my metal arrowheads did not. Using a flint-tipped arrow, I could easily imagine hunting buffalo, but the fantasy just wasn't as rich with commercial metal-tipped arrows.

Civilizations are defined by the materials they use. It is with only a little exaggeration that we speak of the Stone Age, the Bronze Age, the Iron Age. If we had to characterize our own age in material terms, we might refer to it as the Synthetic Age. Yet what is beyond the age of synthetics? We'll come to that. For now, let's develop an appreciation of materials by considering two case studies of materials use, one old and one new.

The first case is an old and complex biological material used by the Plains Indians of North America. It is a complex material that needs to be taken apart, so that uses can be found for the constituents. It is the bison, commonly called the buffalo. The American Bison Association publishes a chart showing the uses of the buffalo by the Plains Indians. No part seems to have been wasted. Here are some of the separate parts that found applications in the inventions and technology of the Plains peoples: hide, horns, skull, brain, teeth, beard, blood, bladder, tendons, meat, muscles, hair, fat, bones, tail, hoofs, chips, stomach contents, stomach liner, paunch liner, liver, scrotum, and gall. To be more specific, bones alone are listed as having these uses: fleshing tools, pipes, knives, arrowheads, shovels, splints, sleds, saddle trees, war clubs, scrapers, quirts, awls, paint brushes, game dice, tableware, toys, and jewelry.

One wonders where the ideas for these applications came from. Certainly, some of them originated in need: given a need, one then looks around for a material that will satisfy it. Others may well have originated by saying to oneself: Materials are scarce; here is a part of the buffalo that I haven't used yet. What can it be used for?

The second case study of materials is a contemporary one and illustrates how diverse materials come together to provide complex functions. Let's examine a quotation from the engineer M. J. French:

> [A] record-player pick-up cartridge may marry a tiny precisely-shaped diamond, a fine strip of bronze three times as tough as anything known a hundred years ago, delicate coils of wire fine as spider's web, a powerful little magnet made of rare metals or oxides, whose existence was unsuspected a century ago, and intricate and perfectly fitting parts of strong plastics and metals. . . . The many different materials of the pick-up cartridge are chosen for a variety of reasons. Some are chosen for special physical properties possessed by very few substances, as in the case of the magnet. The stylus is of diamond because it is the hardest material we have, and the wire is of copper because it is almost the best conductor of electricity, and is moreover ductile and sufficiently strong. The plastics are chosen out of a very wide range of similar materials on a large number of considerations taken together, only some of which have to do with function.

With the buffalo and the phono-cartridge as context, we are ready to impose an organization on this world of materials. The materials of invention can be placed on a rough gradient of alteration imposed on the found world: from natural materials, to the synthesis of natural materials, to the formation of synthetic materials that go beyond nature, to smart materials, and finally to very smart materials.

Natural materials are things like stone and wood. They are there for the taking. Initially, the knowledge of their properties is slow in coming, a matter of incremental trial and error across vast expanses of time. Stone is the oldest material of invention that we know. Its record of use comes to us in stone tools of long-gone ancestors. That record is not a homogeneous one. The evolution of stone tools shows a slowly developing sophistication in which different methods of stoneworking emerge. The trends indicate progressively more control over the working of stone: smaller chips resulting in longer and sharper cutting edges, with less waste of flint.

Stone is also known to us as the material of artistic-religious invention. Stone cairns provided memorial markings over graves, with the earliest prepared burial sites probably traceable to Neaderthal peoples. Carved fertility figurines make their appearance in Cro-Magnon caves. There, stone implements also served as a rudimentary writing surface, with tick marks suggestively indicating a count of long past significant events. Stone was used as a building material of ancient civilizations, and continues that use to the present. And in a triumph of recent marketing, like an ancient superstitious rite, stone came on the modern scene in one of the most unlikely of fads, the pet rock.

Wood as material probably goes back even farther than stone, but because it is subject to decay, its record is ephemeral. Bamboo is an old and important form of wood (strictly speaking, it is a grass). The first historically noted uses of bamboo as a material are attributable to the Chinese, more than 2000 years ago.

Bamboo grows rapidly, is long and tubular with partitions between segments; and its fibers are very long, while being quite strong. Early uses of bamboo were for rope, well casing, and pipe. The search for salt was a powerful motivating force in China, and in noncoastal areas this meant drilling into the earth. The form of drilling was crude by present standards, but the results were not. Bamboo rope was used, with one end attached to a teeter-totterlike lever system for people to jump on and off, and the other end attached to a battering ram bit. As the bit was raised and dropped, a hole was formed. The depths attained were sometimes in the thousands of feet, and this as early as 2100 years ago.

The rope for these drilling rigs was made of long twisted bamboo fibers, and as such was quite strong. On a weight basis, the tensile strength of bamboo approaches that of low-grade steel; unlike hemp, bamboo gets stronger when wet. The casings for these salt wells were also formed from large-diameter bamboo tubes. When carefully fitted together, the bamboo tubes kept the walls of the well hole from collapsing and provided a smooth access for containers to dip out salty waters. On the ground surface, bamboo pipe was used for transport of the salty waters to evaporation tanks, where the salt was concentrated and then collected.

During deep drilling, the discovery of natural gas was inevitable. Probably through cruel accidents of trial and error, the fiery nature of gas was made unknown. In any case, the gas then was piped, by bamboo tubes again, to the evaporation tanks where it was burned beneath the tanks to speed the process of evaporation and salt concentration.

Another particularly interesting use of bamboo is as a model. The compartmentalization of ships began early in China, hundreds of years before the idea began in the West. It has been suggested by the historian of Chinese technology, Joseph Needham, that the compartmentalization of ships came about in analogy to the segmentation of bamboo. If so, this would be an early and effective use of a found-in-nature model. The maritime advantages of compartmentalization are in safety and strength. If a ship's hull is penetrated, only the immediate compartment is flooded, far better than sinking. A compartmental structure also contributes strength without adding much weight, related to the corrugated cardboard boxes we examined earlier. Looked at in another way, this is a miniaturization and parallelism of hulls; its use guarantees that failure in any one hull will not be critical and will not propagate to the others.

Another example of analogy related to bamboo may well be the idea of piping itself. Presumably ditches are a simpler notion than piping. Still, a bamboo pipe offers many advantages over a ditch, ranging from speed of construction to flexibility of placement.

Another early, important, and continuing use of bamboo is for construction. Its light weight and strength contribute to making it a good framework for building and scaffolding.

A more recent use of bamboo shows that it is a material for which new uses are still being found. Edison's search for an incandescent bulb filament provides a worthwhile example. For some time he had been seeking an incandescent filament with certain properties: It should burn bright, but not too bright, for a long time; it should be reasonably strong and reliable; and it should be easy to use in manufacture. After trying a number of materials, he found that for each requirement carbonized bamboo worked quite well. In fact, bamboo was among the earliest useful filaments for the incandescent bulb.

The diverse nature of bamboo's applications indicates the range of uses to which a good material may be applied. But that does not mean that people set out to systematically find uses and determine the properties of materials. Indeed, most of the history of natural materials like stone and wood is one in which a material's properties are determined by the informal experience of trial and error.

A huge step forward in the development of materials is the systematic determination of properties by engineering and scientific studies. What is a material's strength, ductility, melting point, relative hardness, rigidity, and so on? Systematically asking such questions quickly takes us beyond the knowledge acquired by informal experience. We end up with a catalog of materials and properties. When we need certain properties for a task, we can look up the best material.

A next conceptual step on our gradient is to alter a natural material's properties. Adding and deleting properties is a common form of such alteration. We add carbon to iron, in just the right proportion, to produce steel; we add antacid to aspirin to produce a buffered aspirin. We delete impurities from molten iron to provide a tougher and more regular product; we remove caffeine from coffee to produce decaf.

The next step on our gradient is the synthesis of a natural material. In the 1950s, Robert Wentorf, Jr., and his colleagues at General Electric engaged in one of history's great sleight-of-hands. Far surpassing the aspirations of the early alchemists who merely wished to transmute lead into gold, they transmuted peanut butter into diamond. Both peanut butter and diamond are forms of carbon, one of the most abundant elements on our planet. But there the resemblance stops. The atoms of diamond are packed together in a three-dimensional lattice with each atom fixed rigidly to four other atoms. Other sources of carbon such as peanut butter or the graphite of pencil lead have their carbon atoms arranged in sheets of a hexagonal pattern, much like chicken wire. As a result, peanut butter and pencil lead are not very hard, but diamond is the hardest substance known.

Wentorf and his colleagues accomplished their sleight-of-hand by duplicating the natural conditions under which diamond forms: intense heat and pressure from deep within the earth. They designed a special furnace equipped with a piston that pressed downward to subject an ordinary sample of carbon, like peanut butter, to great pressure and heat. After much effort, sophisticated science, and inspired insight, they managed to synthesize diamond from ordinary carbon. Their results were not gem stones but tiny pieces of diamond grit, nonetheless grit with great practical importance for industrial abrasion and grinding. Later, with the right configuration of energy and equipment, they were able to produce small stones approximating gem quality, but very expensive ones given the required energy input.

However, the cost did not matter. What did was the possibility of actually manufacturing diamond. That possibility more recently gave rise to producing *thin films* of diamond on other surfaces, a development of great commercial importance. First, investigators in the Soviet Union, then Japan, and now the United States have managed to coat manufactured objects with these diamond films. This is done in several ways, one of which is through chemical vapor deposition, or CVD as it is called. In a heated vacuum chamber, carbon vapor is produced from a hydrocarbon such as methane gas. The vapor then deposits itself on other cooler items in the chamber, much like ordinary condensation does in a refrigerator.

The process is not easy. To produce pure diamond and avoid ordinary graphite by-products, the temperature has to be very high; and initially the process was inefficient and costly. That precluded coating anything that would melt at the temperatures required and limited the process to all but the most critical metal surfaces.

Recent improvements in the process mean that expensive tools like oil drilling bits and surgical instruments can be diamond coated to greatly extend their wear cycle. Before long, it may be possible to buy a kitchen knife with a diamond-coated edge that will never need sharpening. And Sony is already selling a speaker cone coated with diamond, a cone that is light and rigid, just what is needed for communicating high-frequency sound with a minimum of distortion. Another potential application is the coating of engine parts with diamond film to prevent wear, although the problem of thermal expansion may limit such an application. One of the bigger anticipated applications is the use of diamond semiconductor computer chips. Because diamond is a wonderful conductor of heat, it will be possible to build ever-smaller, faster, and more tightly packed circuits that dissipate heat rapidly. And as the temperature required for vaporization comes down, it will be possible to coat glasses and plastics economically. Perhaps before too long our sunglasses will have a diamond coating to prevent scratching.

The specific applications of diamond synthesis are difficult to predict, but the evaluative trends are becoming clear: an ever lower temperature needed for coating, greater efficiency in the gas-to-film conversion process, coating of ever more materials, and cheaper products that will last "forever."

Let's continue with our materials gradient. We have seen that the ability to synthesize diamond and control its form, from particles to film, is a giant leap in materials science. That's only part of a larger progression. The next step is to synthesize materials not found in nature, materials with properties more extreme or different than anything naturally occurring.

The development path of the plastics and synthetic fibers industries shows the push beyond nature. Initially, the goal was often similar to that for diamond: duplicate properties found in nature. Early attempts in fibers chemistry sought to produce an artificial wool or silk. However, it was not long until researchers wanted more. One of my favorite examples is the invention of Kevlar by Paul Morgan and his colleagues at Du Pont.

Kevlar is a synthetic fiber with unusual properties. It is a polymer which means that the molecules are quite long; in addition, it has high molecular weight. Together this means that extracted fibers are likely to be strong. Indeed, pound for pound Kevlar is much stronger than steel, possessing great tensile strength. In addition, it does not rust, and is a stiff fiber. Here then we have a new material, something that does not exist in nature, and in some respects surpasses anything in nature. It is now used in bullet-resistant vests, for army helmets, and as a lightweight cover for applications that involve hard impacts and abrasion. Additional uses include that of a cable and a possible replacement for the asbestos of brake linings.

The applications took some time in coming. Like any new material Kevlar is a substance in search of applications. Evidently, Du Pont did not have at the time—and may still not have—a group of people with the explicit task of thinking of potential applications. Instead, this thinking process was done quite informally. As we have seen in the related case of the Post-it adhesive, finding applications for materials is something that should have more emphasis; there may be some good heuristics to aid in the process. Notice that finding applications is the reverse of having a computer database listing the properties of materials. In the latter case, we start with a problem, determine the properties of the material that we need for a solution, and then look through the database for something close. The manufacturer of a material should do the opposite as well: find new applications for a material.

Here the two problems—looking for a material with given properties and looking for applications for a given material—are quite different and at least equally important. To my knowledge, the second problem of finding new applications by starting with the material's properties has not been attacked systematically. If we can figure out heuristic procedures for finding new applications for materials—especially first applications for new materials—we will have come close indeed to serving up the freest lunch that technology can provide.

Continuing along our gradient of materials development, we come to *smart materials,* those materials that change their properties to track or counteract changing environmental conditions. A good example is photochromic glass that changes its color in response to light. A sample application is the sunglass lens that changes its darkness in response to the increasing intensity of the sunlight. Such a material is smart in some sense. For human-made materials, an automatic change to accommodate a change in the environment is something new.

While most of the materials used in human invention are unable to change automatically for a shifting environment, biological materials regularly do this. After all, commonplace materials like skin and bone are capable of growing in concert with the rest of the system, self-healing or self-repairing themselves, and certainly self-adjusting to environmental stresses. Skin and bone are *very smart materials.*

A fiber analog to skin is a science-fiction garment that changes its thermal properties to reflect the balance between atmospheric and body temperature, only more flexibly. On a cold day garments made of this fiber should warm us, on a hot day cool us. In feel the garment should be light like silk rather than burdensome like a space suit. In hot climates, it should do the rapid equivalent of tanning, quickly gating out excess radiation and heat. Where parts of it rub on something, then like a worker's hands it should develop the equivalent of calluses to protect against wear. In addition, it should grow with our bodies as we develop, and it should repair itself when damaged. Ideally it will help us—as do some creature's skins—with the mating game by attracting members of the opposite sex at the right time of the year, or better yet at the right time of the day. After all, these are just a few of the intelligent things provided by the real skin of many organisms.

Putting the comparison this boldly shows something of the hidden intelligence of skin and the low IQ of current invented clothing and fibers. However, skin is the product of hundreds of millions of years in nature's R&D lab, and human clothing has been in development for only tens of thousands of years.

I wonder what the next steps will be in intelligent materials? Only the foolish will be specific, but the trend is not hard to discern. It takes millions of years to gain mastery over naturally occurring materials like stone and wood, without any kind of deliberate analysis. It takes a thousand years to develop an adequate materials science, in which the outward properties of materials like steel are mapped with their inner chemical and physical structure. It takes a few hundred years to get somewhat sophisticated about synthesizing naturally occurring materials. It then takes a few decades to develop new materials like nylon with properties not found in nature; then a few more years to begin the development of smart materials like photochromic glass. Human-made very smart materials cannot be far behind. Now if we can just figure out applications for them.

If a civilization is defined by the materials it uses, will we go beyond the Synthetic Age? Might the civilization to come be as smart as some of the new materials now on the drawing boards of inventive minds, materials that aspire to the intelligence of skin and bone?

15

The Interface's Form

An interface is the place and the way that two different systems are connected to one another. A human interface is the way we connect with our artifacts. A typical example is the handle, an interface between a tool head and a human hand. I first got the idea for writing about interfaces by collecting a bag of related metaphors:

- I can't get a handle on this problem. *What kind of handle attaches to problems?*
- Don't fly off the handle. *What part of you is about to come off?*
- Get a handle on yourself. *Where on your body do you want it?*
- Handle yourself with dignity. *Is the handle of dignity in the same place?*
- Get hold of yourself. *What grip do you use here?*
- The dog handler (trainer) was peerless. *What handle and grip did he use?*
- The President's handlers controlled the news conference. *Left for your own interpretation.*

The question beside each metaphor is a bit too literal. What most of these metaphors do is communicate abstractly the idea of controlling the self (one part by another) or controlling another person (a kind of action at a distance). At a sufficiently abstract level, that is precisely what a handle does. Let's now look at the specifics. What kinds of handles go with what tools and functions. What principles of design and interface lie embodied in the match between artifact and handle? Let's begin by looking at some variations on the handle theme; then we'll finish by examining another form of interface, the way a musician comes in contact with a musical instrument.

Getting a Handle on Things

One of invention's rich themes is how artifact and user come together: the idea of an interface. I don't have far to look for examples. The keyboard of my computer is an input interface between me and the information storage and display capabilities of the computer. The display screen is an output interface between the computer and me, what the machine shows of its current state and workings.

A related form of interface, an elemental and seemingly simple one, is a contact point between humans and artifacts—what we call handles. The handles to consider here may be classified in two ways: those concerned with thrust and those concerned with support. The thrust class is exemplified by the knife handle and the support class by the suitcase handle. But each of these has many elaborations, all held together and organized by loose family relations where a minor change in one handle form allows us to slide gradually into another. (I say *slide* because there is no sharp boundary between these different forms, and by changing a few things in one we glide into another.) Here is an invention family with a good deal of structure below the surface; our task is to abstract and organize that structure.

Let's begin with the knife handle. Simple hafting took hundreds of thousands of years to develop. The earliest hominid stone tools, dating to two million years ago, did not have explicit handles. The first tool with a stone working edge and a wooden or hide-wrapped handle is obscure in history, but it held profound advantages over its unhafted ancestors in allowing a better grip and action at a distance. Because the earliest stone knives did not have a handle, let's consider a conceptual account of the handle's development.

An example is shown in Figure 15.1A. This knife was doubtless gripped from above and used in either a slicing, chopping, or scrapping fashion. We will refer to the configuration of the hand as an *open grip.* Notice that with an open grip the full force of the hand, arm, shoulder, and torso is not available. The same is true for the knife in Figure 15.1B.

A better way of gripping the knife is needed. Figure 15.1C shows a top-shouldered blade, a plausible next step. With the top-shouldered blade, *some* of the larger muscles of the upper arm and shoulder come into play. However, the grip is still from above, and it is not possible

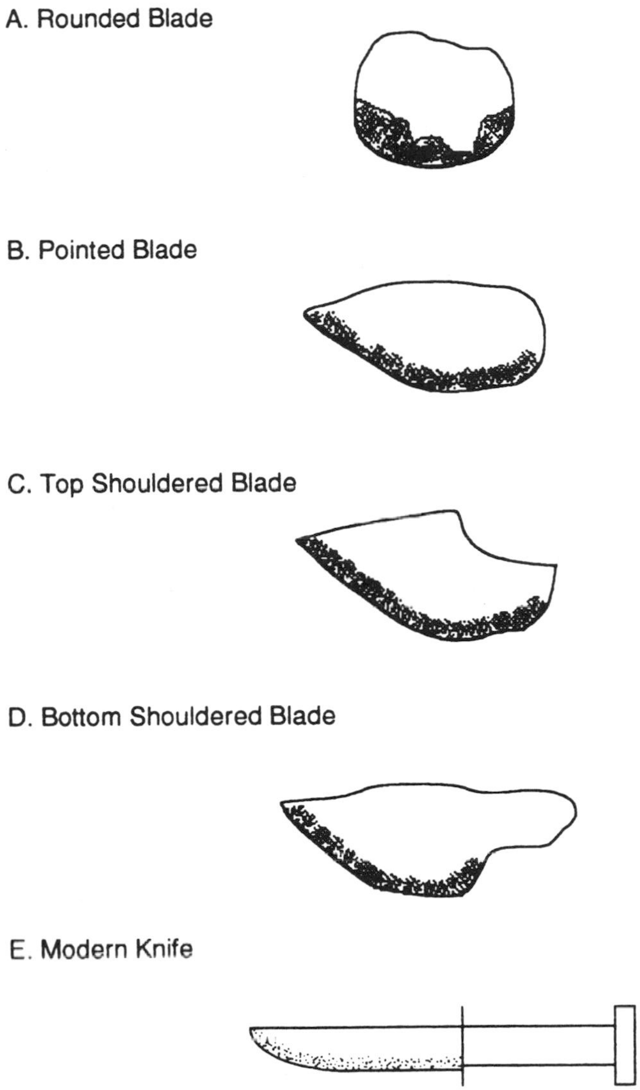

A. Rounded Blade

B. Pointed Blade

C. Top Shouldered Blade

D. Bottom Shouldered Blade

E. Modern Knife

Figure 15.1   The conceptual development of the knife handle. The slow development of an explicit human interface and then its progressive refinement and fine-tuning.

to wrap the entire hand around the handle in a power grip—the way we hold a baseball bat, with thumb and fingers touching. Figure 15.1D shows a bottom-shouldered blade. The bottom-shouldered blade now allows for a complete power grip with the entire hand wrapped around the handle. For the first time, the full array of large muscles in the hand, forearm, upper arm, and shoulder can work in concert.

Problems of fit and comfort remain, however. Even with the bottom-shouldered blade, the hand does not fit precisely the rough edges of chipped stone; the match is poor and comfort is lacking. A solution is afforded by the stone handle around which is wrapped or bonded another material that is more easily formed to the hand, a material like wood or a wrapped leather strip. When that step is taken, we have achieved most of the modern knife's handle form.

Somewhere in this hypothetical progression, emergent capabilities develop. Along with the structural changes in the knife, the grip changes from an open grip, in which the thumb is roughly aligned with the point of the blade but does not close in a power grip, to an underhand power grip that closes about a handle. With the bottom-shouldered blade, the option of an overhand power grip also develops, and the thumb is nearest the butt of the handle, like that used in a stabbing motion or for holding the knife when doing two-handed etching. These new actions and functions develop along with the changing interface between human and tool. That new functions emerge simply by changing the interface to afford a better grip and action is obvious in retrospect, but it is a deep truth that took a long time to reveal itself.

The progression in Figure 15.1 from forms B to C to D suggests other dimensions of interest. We can draw a series of analogies of the form "B is to C as C is to D as D is to ?" and find by extrapolation the developmental path of the knife interface. It is a progression toward more leverage and control, as well as toward greater comfort. Analogies of this form reveal hidden evaluative criteria behind a changing interface.

Now let's look at ax and hammer handles as interfaces. In comparison to an unhafted ax, the one with a handle allows for action at a distance, greater leverage, and increased velocity at the striking head. Thus a handle's distance between the user's body and a target, prey, or a foe increases the user's effectiveness and safety. A handle as a lever also allows for prying and lifting with far greater force than is possible without it. And, for a given rate of arm movement, a handle increases the velocity at the working edge of a tool relative to the hand or elbow at the center of the arc. This is because the tool edge travels a greater arc distance over a given angle with the handle than without it. The hafted tool's greater velocity at impact is directly translated into increased work or effect.

A simple straight handle also allows for continuous adjustment of force and control by moving the grip up or down the handle. The force of the striking tool edge is greatest when the handle is gripped near the butt end, for the reasons just given. However, the control or accuracy in striking is greatest when the handle is gripped near the tool's working edge. By moving the grip up and down the handle, any desired balance can be found between needed force and accuracy. Without the handle, with the arm alone, all the adjustments for force and accuracy are limited by the characteristics of the human arm.

As useful as a simple straight handle is over a tool without a handle, the balance sheet includes a negative. People who regularly use an ax or a hammer (carpenters, roofers), or a tennis racket, often develop pain with an origin in the muscles, tendons, or joints of the arm. A likely cause is the unusual alignment of arm, wrist, and hand with the handle at the time of impact. As shown in Figure 15.2A, when holding a hammer and striking a nail, the wrist at the time of impact is bent downward to keep the grip firmly on the horizontal handle. The problem of a wrist twisted this way is alleviated by having the handle curve as in Figure 15.2B, a design sometimes used with hatchets and axes but rarely with hammers. This same curved design is now used as a cure for tennis elbow. The idea is the same in each case: With the curved handle the wrist can be kept straighter at the time of impact, reducing the distorting stresses in wrist, elbow, and shoulder.

Unlike the hammer, the modern hand saw has a handle that allows for proper alignment of hand, wrist, and shoulder in delivering force. Figure 15.3A shows how a power grip goes through the handle, with the fleshy cushioned part of the hand absorbing some of the thrust. Because the weight of the saw is sufficient for downward force, the entire thrust of the arm pushes the saw forward. To see the rationality of the saw's handle, introduce a test for design: What if the opening in the handle were parallel with the saw blade, as in Figure 15.3B? If

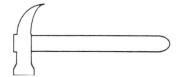

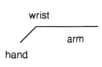

A. Claw hammer, with a straight handle.   The arm/wrist angle at impact.

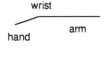

B. Hammer with a curved handle, like that of an ax.   The arm/wrist angle at impact.

Figure 15.2   Straight and curved handles. Variations in handle shape and wrist angle at impact.

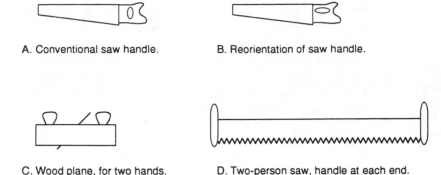

A. Conventional saw handle.   B. Reorientation of saw handle.

C. Wood plane, for two hands.   D. Two-person saw, handle at each end.

Figure 15.3   Handle forms. (A) is a conventional handle; (B) the same handle form rotated; (C) is a two-handed handle on a wood plane, and (D) is a multi-handed handle for two people, as on a two-person saw.

the arm operates with a pistonlike stroke, the result is an exaggerated bent wrist action that we observed with the straight hammer handle. Naturally, I am assuming here that the arm works with a pistonlike action. If sawing is done with a swinging arm action, as in Japan, then Figure 15.3B may be more appropriate. Again, the fit between grip and action is an important idea.

The wood plane's handle exhibits yet a new conceptual form, the two-handed handle. Figure 15.3C indicates the two handles, where the left hand is placed on one and the right on the other. Using two hands instead of one supplies greater thrust (both hands and the body behind the action) and added control (each hand contributes to linear forward movement and prohibits skidding to the side).

We have moved from the one-handed handle to the two-handed handle. A continuing line of generalization is to handles requiring even more hands, a feat accomplished by bringing in other people. Figure 15.3D shows a two-person saw, a conceptual slide into yet another handle form. Each person may use either one or two hands, with heavier work more likely to require the two hands of each person. The design allows for a power grip with each hand and a maximum thrust from arms, shoulders, and body, a joint parallelism of several hands and several users.

The next class of handles provides suspension and support rather than thrust. The prototype is that of the suitcase with the handle over the container's approximate center of gravity. A comfortable power grip is wrapped around the handle, and the container suitcase is suspended on the side close to the body. The handle as human interface and the suitcase as container come together to create a portable package that can be held in one hand.

To see why this package of handle and container is important, try eliminating the parts and noting the effects. In the absence of handle and case, items like clothes must be bunched together in both hands and arms, with the hands and arms acting simultaneously as container and support; many muscle groups are involved. Now add a container, like a cardboard box without a handle. The container alone requires that both arms be slid under it for support. It is held in front of the body because both arms are necessary, as is holding it close to the body's center of gravity. When we add the containment afforded by the box, a much larger group of clothes can be held together. Finally, we add a handle to the container to form a suitcase. Now the suitcase is held with one hand at the side close to the body's center of gravity. The load hangs from the skeleton rather than being supported by the muscles (fewer muscle groups are involved), and therefore greater weight can be carried.

To further see the workings of design here, shift the handle around. When the handle is rotated ninety degrees on the suitcase so it is parallel to the left right axis of the body, the arm must be twisted to carry the suitcase at the side. Alternatively, when the handle is moved from the center of the suitcase to its corner, everything is off center and out of alignment with the likely center of gravity. We can try other variations also by changing the handle's size. To fit the hand, a handle must be made in a narrow size range. Again, we find that hidden intelligence is present in the relation between a suitcase and its handle. Too small a handle and the entire hand will not fit through it; too thick and a power grip will not wrap around it.

The yoke and dual water buckets shown in Figure 15.4 (top) allow another conceptual slide into an important form, the no-hands handle. The yoke over the shoulders allows support over the body's center of gravity. The two buckets balance one another, and for the skilled carrier, both hands are free. Should this arrangement be called a *handle?* It is certainly debatable. The justification is that the water-bucket yoke has many characteristics in common with unambiguous forms like the suitcase handle, including a common function (the suspension of loads) and a common mode of conveyance (by the body). Seen in this light, the water-bucket yoke is a legitimate case of a zero or null handle that interfaces not with the hand but with the shoulder.

Generalizing, we can see that the carrying pole in Figure 15.4 (bottom) is a two-person yoke requiring no hands. For reasons similar to those pointed to in the case of the water bucket handle, I will also classify this device as a handle, even though no hands are used. In simple form, it embodies the important principle of parallelism—this time among people and achieved by connected or cooperative action between individuals.

When the idea of the handle is highly abstracted, one or more of these characteristics is found important: conveyance of thrust, action

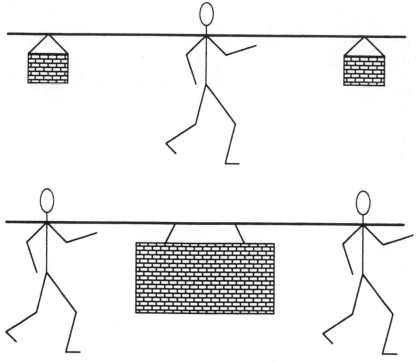

Figure 15.4 Water bucket poles. One-person and two-person poles, with null handles. Both hands are free in both cases; the freedom allows for better balance, or the capability for carrying other things.

at a distance, suspension or support, and above all a connection between different kinds of systems. In short, an interface.

Another different kind of interface, with many comparable characteristics, is of paramount historical importance. It is the way of connecting human control and animal power. Good examples of this are the ways humans use horses for riding and for pulling. The respective interfaces that underlie each of these actions are the stirrup and the harness.

The stirrup is a form of "handle" in that it supports a load (like the suitcase), but in this case the load is oneself. Also, it enables the conveyance of thrust, either for the knees to steer the horse or for the hand to support a lance. However, the load is now *above* the handle and pushing down on it. Together handle, load, and thrust have slid into a different geometry, illustrating once more that invention categories do not have sharp boundaries.

The historian of technology L. White describes the fascinating story of the stirrup, a device that revolutionized medieval warfare. Without the stirrup, staying on the back of a running horse requires great skill in balance while holding onto the horse's mane. With the stirrup, much less skill is needed, both hands are freed part of the time, and those hands may then hold heavy shields and weapons. The stirrup makes the deployment of heavy calvary practical. For the first time, a rider coated in mail, with lance or sword in one hand and shield in the other,can keep his balance on a charging horse. Without the stirrup, calvary and knights could not have played out their considerable role in the battlefields of history. Horse calvary and knights held sway until the extensive use of gunpowder, although a well-placed bolt from a crossbow could also penetrate much of the armour.

A far older and more peaceful interface than the stirrup is the horse harness. This harness conveys thrust and pulls a load, so in some respects it overlaps with a handle. But it is no longer a human interface; instead, it is an interface between horse and cart. Before the third century B.C., the Chinese were using the harness illustrated in Figure 15.5 (top). It is a configuration that fits around the horse's neck, and we shall call it the *throat collar*. With the throat collar a horse can pull a maximum load of approximately 550 pounds; any more and the horse begins to choke. In contrast, the *shoulder collar* of Figure 15.5 (bottom) allows a horse to pull approximately 3000 pounds. This vastly improved performance is possible by simply changing the interface to apply the load across the horse's chest rather than windpipe. The transition from the throat collar to the chest collar took China hundreds of years, and a thousand years elapsed from the first Chinese use of the shoulder collar until Europeans began its use.

The advantage of the shoulder collar is obvious, so why did it take so long in coming? Suppose you had to choose between your own human efforts at pulling a plow and those of a horse pulling it with a neck collar. The five-hundred pound thrust of the horse makes your own efforts puny in comparison. Would you immediately look for a better way, or be grateful for your liberation? The idea of continuously improving the artifacts around us must be slow in germination and narrow in generalization, when it does occur. That the improvement from throat collar to shoulder collar took hundreds of years is therefore not as surprising as the idea that an invention can always be better. As children of the twentieth century, we have a modest grasp of the idea of continuous improvement applied to our invented technology; but for our invented social institutions and traditions we are much like the ancient Chinese and later Europeans who probably were delighted with the neck collar. Undoubtedly, later generations will look upon our social inventions with the same unbelieving stare we fix on the neck collar.

A contemporary example illustrates the subtle nature of continuous improvement as an idea. I sit here and reflect on how satisfied I am with a modern computer for my word processing. After all, con-

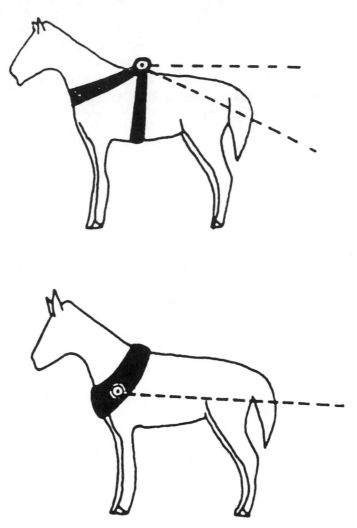

Figure 15.5 How to interface a horse and wagon, the neck collar (top) and the shoulder collar (bottom). (From J. Needham. *Science and civilization in China*, Vol 4, Part II, p. 305; by permission.)

sider the alternatives: typing on an old typewriter; or, worse yet, writing with a lowly pencil. Yet the present interface between human and computer, known as a keyboard, has its own disadvantages. If you were in the room with me, I could tell you what I wanted to say in a fraction of the time it takes me to type the same information. I realize this because I live in an age in which change is more familiar than constancy. To live in a simpler time, with slower moving processes and limited communication, might well have made me content to bear uncomplainingly the lifetime yoke of the keyboard. Now I sense the possibility of computer entry by direct speech, an interface somewhere just beyond the image of the screen, perhaps a hundred essays away.

A Place Between Hand and Sound

Let's expand on the idea of an interface between human and device. The way we relate to and stroke resonant objects governs the nature of the emanating sounds, sometimes merely a scratchy noise and at other times glorious music. While the ways to make objects resonate are very diverse, ever more of today's music is produced by instruments with keyboards. This is particularly true of popular music, the music of commercials, and street music. Why? And what will the answer tell us about invention?

To begin with, the keyboard itself is not a musical instrument. This is evident in the many mechanisms of sound production linked to the keyboard. One keyboard mechanism plucks the strings of a harpsichord, another strikes the strings of a piano, and still another opens and closes the air valves of a pipe organ. What then is the keyboard? The keyboard is an interface between a player and a mechanism of sound production. It controls or dictates what a human must do to extract sounds from an instrument, regardless of how those sounds are ultimately produced.

Historically, except for the organ, instruments with a keyboard interface produced sounds that rapidly damp and have minimal variation in timbre. *Damping* is important for the form of the sound waves. It determines how long the sound lasts and how it trails off; a piano's notes damp rapidly, a bell's slowly. *Timbre* refers to a characteristic sound quality. It is timbre that distinguishes a piano and a violin playing the same note. Technically, a complex sound consists of a fundamental frequency and a series of overtones. The relative strength and pattern of those overtones distinguishes the timbre of the piano, violin, or other instruments such as a trumpet.

Of course, the keyboard is not the only musical interface. In the violin family a bow is pulled across strings, and it allows for a broad range of damping and adjustments of timbre. By controlling the bow's velocity, its closeness to the bridge, and the pressure applied to it, sounds of any reasonable duration and with many timbres are produced.

When we compare the piano and the violin, we find that each interface and each mechanism of sound production has advantages and disadvantages. The comparison of the keyboard and the bow reveals the characteristics of sound controlled by each interface and an associated method of sound production. The piano's keyboard has two great advantages over the violin's bow: parallelism of voices and error tolerance for finger placement. Parallelism here means that many voices or parts can be produced at once by a single player. The piano has as many as ten simultaneous voices, one for each finger. The violin is particularly deficient in parallelism. It cannot produce more than two simultaneous voices (using broken chords, one may get up to four, but they cannot really be called simultaneous). The string placement over a rounded bridge and fingerboard and the requirement of bowing selected strings all limit the violin's simultaneous voices to two. The keyboard interface emerges as the winner when the criterion is parallelism in voice production.

But how tolerant of an error in finger-placement are the interfaces of the two instruments? A brief digression is in order. The piano's interface is the same for both hands, a keyboard. For simplicity we have talked of the violin as an instrument played with a bow, but actually it is a bowed *and* fingered instrument. Its interface differs for the two hands. While the bow is held in the right hand, the left hand is doing something completely different. The left hand moves up and down the neck, and the fingers alternately fall and lift from the fingerboard to control the length of the string and the resulting pitch.

Under these circumstances, the species of pitch errors produced by the violin and piano interfaces will differ. With the keyboard a wide error tolerance in finger placement is allowed before disturbing pitch. As long as the finger is somewhere over the target key and not over an adjacent key, the right note will sound. When error occurs, it is a discrete pitch step and easily detected by the ear. Short of error, tactile feedback tells one that a finger is near the edge or middle of a key. This contrasts with the errors produced by a bowed and fingered instrument like the violin, where extreme care must be taken in the placement of the fingers or bad intonation will result. When error in finger placement does result, the feedback is primarily to the ear, since few fine-grained tactile cues are present on the continuous neck of the violin.

The piano's parallelism and error tolerance combine so one can soon learn to play basic musical compositions. The violin, however, requires extensive practice to master the fundamentals of correct intonation. The ease with which children play rudimentary melodies on the piano and the considerable difficulty they have in playing the same melodies on a violin illustrate clearly that the keyboard is again the winner for learning speed.

As a former violinist, I would like to find at least some properties that give instruments played with a bow an advantage over instruments played with a keyboard. Are there such properties? Yes, and they concern the degree of control expressed over the sound wave for the variables loudness, pitch, and timbre.

The control of loudness on the piano, disregarding the limited functions of the foot pedals, is determined by how hard the keys are struck—the attack parameter. Besides attack, the performer has limited control over the rest of the sound wave envelope. For example, loudness after the attack cannot be increased. The violin, however, gives the player exquisite control over attack, sustain, and release through bowing acceleration, speed, and pressure. The entire sound wave envelope can be shaped at will, although it takes considerable skill to do this artistically.

The two instruments also differ in their ability to produce delicate pitch variations. Each key of the piano keyboard is fixed in pitch. All scales are well tempered—that is, all the adjacent note intervals have the same frequency ratios. This is a compromise that allows the piano to be played in any key, although the subtle tonal colors of other methods of tuning and deliberately shaded pitch relations are lost. With the violin though, anything is possible for pitch, because the neck of the violin is continuous and the placement of the fingers continuously variable. The curse of intonation difficulties for the beginning violinist becomes the blessing of unlimited artistic manipulation and nuance for the master performer. Although great skill is required to take advantage of deliberate pitch variations for interpretive reasons, the violin is superior to the piano in this respect.

Finally, we come to the control of timbre, those variations in wave complexity that produce the characteristic sound quality of the trumpet, violin, flute, or piano. In timbre control, the piano with its keyboard interface is severely limited. Its strings are struck with a hammer via the keyboard. While the timbre does change with the attack, the differences are small, nor are those differences greatly altered by activation of the pedals. In contrast, the violin with its bow interface allows for more variation in timbre. Changing the bow's downward force and duration as well as proximity to the bridge allows for a variety of timbre textures. Plucking the strings (pizzicato with the right hand) and the production of vibrato by the left hand are two additional ways of changing the violin's sound quality.

For controlling the sound wave and its envelope, the violin's bow wins over the piano's keyboard. Still, there is nothing in either left- or right-hand manipulations of the violin that produce timbre changes comparable to that between different instruments. Ideally, we may wish for the timbres of the trumpet, flute, violin, and piano. And we want them all in one instrument, whatever that instrument might be.

Why not combine the advantages of both the violin and the piano interfaces in the same instrument, that is, join the parallelism and error control of a keyboard interface and the control over the sound wave envelope afforded by a bow? An additional wish for our list is that the combination allow an even wider range in timbre than now produced by single instruments, a range equivalent to that of different instruments all combined or joined together.

Some modern synthesizers aim to do just that: combine the advantages of the keyboard interface, the violin's sound wave control, and the capability of selecting an "instrument" or, more precisely, its timbre. It is an ultimate complementarity of strengths, the most desired characteristics of each instrument. The success of a synthesizer in producing quality sound is the subject of controversy. But there is little doubt about the synthesizer's direction of movement as it stretches octopus-like to grab the attributes of different instruments, to form a kind of super instrument in a grand synthesis—or in pale imitation, if you don't like its sound quality.

Let's examine a hypothetical super synthesizer, one that does not actually exist but is a composite of real synthesizers and experimental synthesizers. This ideal synthesizer has a keyboard interface, but with a difference. The velocity with which the keys are struck produces variations in attack, the audible aspect of the initial part of the sound wave. Varying the attack with changes in velocity is just the electronic counterpart of varying the striking force on the piano keyboard or varying the rapidity of movement and pressure as the violin bow is pulled across the strings. The keys of our super synthesizer are also pressure sensitive, so increasing downward pressure on a key, *after* it is depressed, will increase the loudness of an *individual* note. The piano keyboard does not have this function, but through the increase of pressure on the bow during a sustained note the violin does.

In our hypothetical synthesizer, each individual key is also pitch sensitive. If a key is pushed forward after being struck, the pitch can be "pushed up" or raised, and if it is pulled back toward the body, the pitch is "pulled down" or lowered. The piano keyboard has no counterpart to this function, but it is readily accomplished with the violin through the movement of the fingers on the neck. The violin's control of pitch is still much more sensitive, continuous, and natural than pushing and pulling of the hypothetical synthesizer's keys, although such a synthesizer has a more controllable pitch than the piano.

All the manipulations described here are to be independent for each key. One key may receive a vigorous attack and a gradual release; the key next to it may receive the reverse, a light attack and a gradual buildup of loudness; the two keys may also enjoy independent pitch variations, one raised by a partial step and the other lowered. Artistic use of the additional control over the sound wave envelope will require learning, but that is what technique is about.

Our ideal synthesizer should also have a range of timbres: sounds characteristic of the piano, violin, bells, organ, and other instruments. Independent timbre control for each finger may not be achievable; however, it is possible for each hand. Just split the keyboard at an arbitrary place, set the appropriate octave range for each split, and then produce, say, a trumpet line with the left hand and a simultaneous flute obbligato with the right hand. How can these timbres and keyboard splits be accomplished? Certainly, a number of switches set easily by one hand can be placed above the keyboard. Since we are now controlling the sound wave envelope by key velocity and pressure, we no longer need the traditional piano's foot pedals for damping or sustaining. In our super synthesizer the foot pedals are used to activate the more common keyboard splits and timbre settings. Ideally, the user will program the various foot pedals for the desired settings.

Our synthesizer should have even more timbre control capabilities. We can insert a tape or integrated circuit memory for new timbres. An organist, for example, may wish to have a library of timbres corresponding to those of famous organs. An avant garde composer may wish a library of nature sounds or their combinations.

Yet there is still more to the development of our super synthesizer. People are now beginning to join the synthesizer with the computer. This is a marriage of complex families that brings new power to both, as well as in-law problems. Again, rather than discussing a particular computer and program, let's choose an abstraction of many.

The advantages of joining the synthesizer and computer are many. One such advantage is the easy entry and editing of a musical score. A voice or part is entered into the computer memory by playing the synthesizer keyboard, or it is entered note by note on the computer screen, by using a menu and a mouse. Once a voice is entered into the computer's memory, it can be easily edited by changing particular notes on the staff, electronically cutting out one part of the score and pasting it at another point, or copying a part to several other places.

A second advantage is instant playback. Once a score is entered, we can play it back either through the synthesizer or the computer's native speaker. Each part or voice is played back individually, or all voices are played at once.

A third gain is the ability to rapidly change dynamics. Dynamic changes are adjustable and programmable. Many computer programs for composition allow the user to make arbitrarily large dynamic changes, simply by inserting the appropriate command in the score. If desired, this is done on a note-by-note basis. Simple edit commands make it easy to change dynamics.

A fourth and very impressive accomplishment is the capability of adjusting timbre. A selected timbre is assigned to the entire composition, to an individual voice (instrument), or to individual notes. With a change in timbre, a given piece is played back as a string quartet, a brass quartet, a piano work, or as a combination of violin, piano, guitar, and organ—if we really want it. The ability to make timbre changes will be invaluable for the budding composer learning orchestration. Because changes can be made so rapidly, the aspiring composer can immediately try out new instrument combinations.

Fifth, the score can be instantly printed. Once the score is edited and is satisfactory in computer performance, the entire score is printed out on paper for use with standard instruments. Or the parts for individual instruments are printed out so the time-consuming copying of parts is no longer necessary.

Finally, it is possible to digitize naturally occurring sounds. Several inexpensive sound digitizers are now on the market. These devices are largely amusing toys, but the same technique used by a moderately powerful computer is another matter. A sound of one's choosing, perhaps a frog croaking or a door squeaking or a bell striking, is fed into a microphone and passed on to a digitizer. There the sound is sliced into brief temporal intervals at a given sampling rate, and each interval is given a mathematical or digital description. The sound then is played back, but because it is based on a mathematical description, its pitch and loudness can be readily varied while its distinctive characteristic sound or timbre is preserved.

We have only made small scratches on the canvas of possibility. Through new interfaces and computer power, the contemporary composer is beginning to have a pallet of sounds available that may be shaped to perfection, that may be combined in ways barely imagined a decade ago—and all this done quickly and efficiently. The nature of music production and composition is changing at breakneck speed. The passing landscape is so blurry that most of us are not even clear on the direction of change, and certainly we cannot begin to predict a destination.

The interface that connects inventions and people, or that connects different systems to one another, is a subject worthy of much more study. Our discussion of handles and musical interfaces is little more than a sampler. Often an interface makes the difference between a usable, accessible product and something shunned.

16

The Art of Containment

I'm looking at a tea bag. It contains ground up tea and has a string handle. Not all the secrets of invention are hidden here, but we can abstract more than one important idea. Indeed, a very important class of inventions consists of those forms that contain or hold things. Let's examine some underlying principles of containment and holding inventions. As we discover and abstract these principles, we will find important generalizations that show surprising relations and linkages among forms of containment. We begin with tea bags, move to a cooking pot and generalizations on it, then to clothing storage and retrieval, and finally to highly sophisticated forms of containment, the cartridge and prefabrication.

The Tea Bag Ceremony

The tea bag I was staring at is a Lipton "FLO-THRU" bag; a schematic version is shown in Figure 16.1. It comes packaged in a small envelope. The envelope is one of a number in a box covered with a cellophane wrap. To get a tea bag out of its individual envelope requires tearing loose a perforated square to which a string is stapled. The string is about five inches long, and its other end is stapled to the tea bag. The tea bag itself is a *W*-fold configuration of filtering paper that is stapled at the top to the string.

Because the tea bag is both handle and container, it is a fine tran-

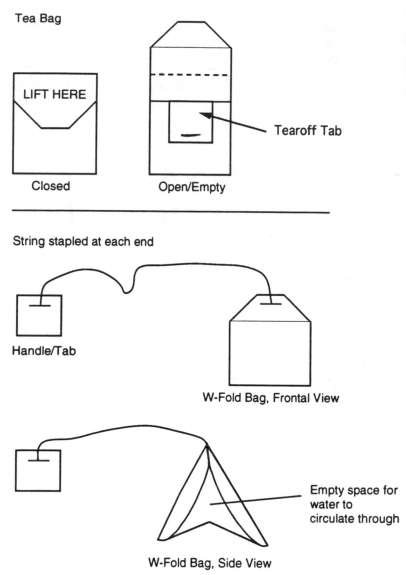

Figure 16.1 The anatomy of a tea bag and its envelope.

sition between the previous chapter and this one. Important invention principles are brewed in the transition. To see these principles most clearly, we need to examine an ancestor of the tea bag, the teapot. In the most simple pot, bulk tea is measured out, heated water is poured over the loose tea so it may simmer, and the resulting brew is strained either before or at serving time. Several subprocedures now come into play. A sieve is held over each cup as the tea is poured—not too elegant and maybe even messy. Or, as an improvement, we use a special teapot with holes at the base of the pouring spout to filter out the leaves. Sometimes the pouring spout gets plugged up; and even if it works during serving, we must still clean the tea sludge out of the pot afterwards. And both the auxiliary and the built-in sieves require separate additional operations, heating the water in a kettle and then pouring it into the teapot. Finally, notice that with the teapot, one strength of brew fits all.

The tea bag offers numerous shortcuts and advantages over the teapot. It minimizes the number of operations required to prepare, serve, and clean up after drinking tea. No measurement of tea is required. It is like a cartridge; a measured "charge" is included in each bag or *container*. The water is still heated in a kettle, but each person has an individual bag and may control the desired strength of brew by varying the time the bag is in the cup. After a suitable time, the bag is lifted from the water by pulling on its *handle*, the paper tab and string. Later, holding it by the handle, the bag is conveniently tossed in the garbage, and the cup is easily cleaned because no leaves have stuck to its sides.

Let's now examine some important components of the tea bag system, as revealed by the contents of my kitchen cabinet. The system consists of these parts:

- *A box covered with cellophane.* A group (48) of envelopes (containers) is encapsulated or held together in one lot or purchase package. The cellophane-covered box is a super container that keeps out moisture and dirt.

- *An envelope.* A single tea bag is encapsulated in one unit (another smaller container).

- *A paper tab.* A paper tab (handle) comes from part of the encasing envelope. The tab handle is used as an interface between tea bag and user. The tab allows for easy pickup, we can drop the bag into hot water or remove it from the water.

- *A string.* An interface or connecting link between handle tab and tea bag, the string is stapled at one end to the paper handle tab and at the other end to the bag.

- *A bag.* The container for finely ground tea, the bag is made of filter paper that allows water *in* to mix with the tea and allows the fluid mixture *out* into the cup, while containing the solid particles of tea. The bag then is a selective filter. Its selectivity is furthered by the *W*-fold shape that allows a large surface area for the water to circulate around and through.

The various design principles of the tea bag are shared with other inventions. Packaging provides individual, transportable modules; the levels of modularity extend from the aggregate box to the individual serving or envelope. Related to the pocket knife, in which the case is the *same* as the handle, the tea envelope contains the paper tab handle as a *part* of the envelope case. Like the pocket knife's handle, the tea bag's handle allows for safe action at a distance: By gripping the tab handle, one can safely move the tea bag into and out of the hot water. In addition, the tab handle is more sanitary and less messy than jiggling or picking the bag out of the water with the fingers.

The tea bag's handle differs from most other common handles in being very flexible. Other flexible handles do occur, however, and include those of the lariate and the whip. Indeed, I found another tool, far more sanguine, that bears some resemblance to the handle and the tea bag. It is the *flail*, a weapon used by medieval knights, which consists of a chain handle with a spiked ball on the end. The flail is a formidable weapon that allows one to swing the spiked ball around an adversary's protective shield. Like the tea bag, the flexible handle provides action at a distance and around barriers. For the flail the hostile environment is one's opponent. For the tea bag it is the hot water.

Taken together, the bag, string, and tab handle of the tea bag function as a disposable cartridge that is fired once, with the waste easily removed. Prefabrication is importantly involved, as it is with the cartridge idea. The prefabricated case has the usual functions: To bind together distinct components into a spatial unit that is separate from other objects.

All told, the tea bag system is a neat one of nested containers and a convenient handle. But the idea of containment goes far beyond the tea bag.

Sliding In and Out of Containment

Earlier, we discussed James Wollin's double-lidded peanut butter jar as an award winner in the *Weekly Reader* contest. As James expressed it, "People can reach the food at the bottom of the jar simply by opening the bottom lid. There is no longer any food waste." I have found myself fascinated with the double-lidded peanut butter jar.

Let's generalize. Is it useful to have three or more lids? Zero lids? Why do we need lids? What are the precursors of lids? Where will these questions take us, as we slide the ideas of lid and containment from one invention incarnation into another?

In fact, lids presuppose containers, so we really want to begin with containers, while working in lids on the side. A biological model of a container is provided by the cupped human hand, and the body itself is a container of the foods and liquids we eat and drink. Possibly, the earliest invented containers were animal parts, such as stomachs, bladders, or pelts. More recently crafted containers include boxes (wood, cardboard), bags (cloth, paper), suitcases (hard, soft), and, at a larger scale, architectural entities like rooms and dwellings. At a more abstract level some containers do not even have walls: plasma gas containers for holding charged particles in a magnetic field; high-energy sound waves that generate "caves" of pressure, where pure chemical compounds can be mixed without contact with contaminating container walls.

What I want to do now is to show how the container idea can be abstracted, generalized, and transformed in a variety of ways. The resulting container forms reveal surprising close family relationships to one another.

Let's begin with a prototypic container, a cooking pot. The cooking pot holds a raw material, solid or fluid food, that is to be processed by heat. The functions of the pot are several: Encapsulation and storage of food, a filter through which heat from a flame passes (but notice that the flame itself does not pass through, only the heat). Therefore, the pot is also a *selective* filter: flame = No; heat = Yes. If we wish to contain the heat after it enters the cooking pot, rather than having it pass through, we may add a lid on the top.

The cooking pot is a three-dimensional container, open at the top if there is no lid. This means that its contents are spread out along three axes: Length and Width are the horizontal axes of the floor, and the third axis is a vertical up and down one, essentially the Height of the pot. For narrative convenience, suppose that the pot is roughly square, as are some electric frying pans. Then the food in the pan is contained in a space that has Length, Width, and Height; that is why we say the pot is three dimensional.

While the cooking pot is a three-dimensional container in this sense, there is another sense in which it is not. Without a tightly sealed lid, some rotations of the pot will leave the food contained, and others will result in it spilling or emptying. Any rotation about the vertical axis is fine; no spillage results. But a rotation of the pot about either of the horizontal axes will spill or empty the food. Notice now the asymmetry between emptying the pot and filling it: We fill from the top (pour things in) and empty by rotating about the length or width axis. Of course, we can also have a more symmetrical emptying from the top by using a ladle, which serves as a smaller temporary container during transport. When a really tight lid is used, the pot may be oriented in any direction without danger of spilling, a useful state if we are transporting its contents.

Remember that a common route to ideas or their organization is to generalize on dimensionality. Do other containers have a different dimensionality than three? Yes, from a practical view, a plate is a two-dimensional container. We might use soup as our test for three dimensionality. Any container lacking the ability to store a liquid, like soup, we will say is at most a two-dimensional one. Like the cooking pot, the plate may be *moved along* the three axes without spillage or emptying, but any rotation about either the length or width axis results in spillage. Incidentally, the table that the plate is on is also a two-dimensional container. Items may be moved over its plane, but the table will not hold soup. Of course, it will hold a puddle here or there because of surface irregularity, but we are talking about idealizations here.

Let's continue to generalize on the dimensionality of containers. So far, we have identified three-dimensional and two-dimensional containers. What about a one-dimensional container? Does it exist? Yes, a track such as that of a shower curtain, a clothes poll for hangers, or a cable for a ski lift or an overhead tram—all these are essentially one-dimensional containers: only to-and-fro movement occurs along the track. For example, the shower curtain and the clothes hangers may be moved along the length axis, to and from, but not along the width or height axis. Of course, with the clothes poll the hangers can be taken off, but then we are no longer in the track's universe of operation.

We return now to the three-dimensional cooking pot. Tacit assumptions were made: The material in the cooking pot was subject to the forces of gravity and was heavier than air, a material something like solid food or soup. Under these circumstances, the material to be cooked gets into the pot from the top and is removed from the pot by either tilting the pot or ladling it out from the top. This means that any openings in the bottom of the pot will produce leaks.

But what if the material in the container is lighter than atmospheric air? Then openings on the bottom of the container are perfectly acceptable, because nothing that is already in the container will run out, and any new material put in the container will have to be *put in from the bottom* to keep it from spilling (this is not a typo). Then to empty the container, an opening will be required *at the top.*

Impossible? Not at all. In some form, this is the insight that first made possible human flight in the hot-air balloon of the Montgolfier brothers in 1783. This was one of invention's great moments, and it was founded on error. The Montgolfiers knew hydrogen was lighter than air (from Cavendish's experiments), and they thought that burning straw would produce hydrogen. They were wrong about the origin of hydrogen but right about the behavior of hot air. Hot air rises precisely because it is lighter than air at atmospheric temperature. If hot air enters the balloon, the container, it will provide lift. As additional hot air generated from burning straw flows up into the balloon, the lift increases until one is airborne.

Once aloft, how to descend? Following the cooking-pot model for emptying contents, we might ladle the hot air out the bottom of the balloon, since that is where it enters, or we may tilt the balloon upside down so the hot air will *flow up*, and out (remember that hot air is lighter than atmospheric air). Neither of these operations is satisfactory for the balloonist. Ladling hot air is an operation more metaphorically suited to a political candidate than a balloonist. And the balloonist can hardly tilt the balloon upside down, because he or she is riding in a gondola at the bottom of the balloon. To do so, the balloonist must commit an impossible act: step outside the gondola, find some place to stand in midair, and then tip the balloon. How then to rid the balloon of hot air? To return to the cooking pot analogy, remember that if holes are poked in the *bottom* of the pot, the contents empty out without tilting the pot or ladling it from the opening in the top. Now since the hot-air balloon and its contents are complements to the cooking pot and its contents, why not have holes in the *top* of the balloon to let the hot air out? That is precisely what is done. A rope-controlled valve (lid, if you prefer) is pulled down *into* the balloon, and the hot air, which naturally rises, escapes from the top of the balloon by *spilling up* the hole. Then descent begins.

Our concept of the hot-air balloon uses the same insight as James Wollin's for the two-opening peanut butter jar. An opening at one end of the balloon lets hot air in, providing lift, and an opening at the top lets the hot air out so we may descend. Next time we see a peanut butter jar, we should think of James, and then mentally transport ourselves to the exhilarating image of being aloft in a hot-air peanut butter jar.

Let's generalize now in other directions. Lids on cooking pots are typically "all or none," that is, on or off. Sometimes this is not satisfactory because things will boil over, so we slip the lid to the side a bit. By so doing, we are sliding the function of the lid into that of a continuously adjustable valve. A more elaborate and intentional valve is a spigot like a water tap. Its operation varies continuously from completely closed to completely open. Should a cooking pot's lid be redesigned to function explicitly as a valve? Some lids have an adjustable steam vent, as on electric frying pans. Pressure-cooker valves are also adjustable.

We can follow yet another path of generalization. A typical container has one lid, and James's had two. Suppose the limiting case. Is there a container that is *all lid*, one in which each side of the container opens? We might call this a *max lid*. We come close to the max lid with jello molds and also with the multipart forms that are used for intricate metal or ceramic castings. When the liquid in a form has set or solidified, we then disassemble the entire container—in effect treating it as if it is a collection of lids. Here we have talked of a max lid-container for the state transition from liquid to solid. Generalize a little more. Are there other situations in which a state of matter, a liquid or a gas, can be "cast" or shaped and then released by varying lids or valves? If so, what uses would such a device have? An interesting thought problem.

Instead of solving it, we want to dwell on another direction of generalization. James's peanut butter jar and the hot-air balloon both had their openings and lids at the top and bottom. Can we think of any containers that have their opening or lids on the *side?* Yes. Many do. Cabinets, doors to rooms and houses—all these are essentially containers with hinged lids (doors) that open on the side.

Another container with an opening on the side is shown in Figure 16.2. It is a fish trap, and it typically has a funnel-like side opening that directs fish into the larger container or trap. In addition to the container idea, a filter is again present. Water passes through the netting, as do fish and other creatures too small to be of interest. But why is the opening to the container-trap on the side? Because of fish behavior; they spend more time swimming in a horizontal or lateral plane

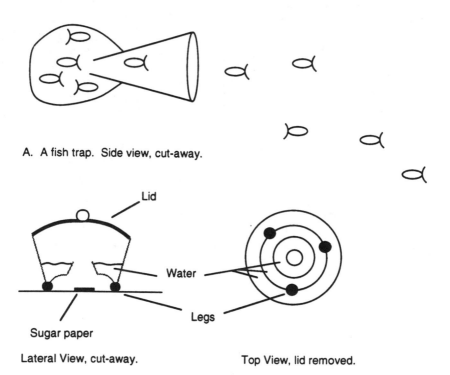

A. A fish trap. Side view, cut-away.

Lateral View, cut-away.   Top View, lid removed.

B. A mystery container.

Figure 16.2 (A) A fish trap as an early sophisticated variation on the idea of containment, and (B) A mystery container. Try to figure out what (B) is used for. Let me know what you come up with.

than in a vertical plane. Hence they are much more likely to enter a trap with the opening on the side.

The fish trap offers another contrast to the cooking pot and the hot-air balloon. The funnel-like opening to the trap is a generalization on the lid concept. A lid need not be simply On or Off. As noted, at a more general level, a lid is a valve that when open will admit some things and when closed will keep other things from coming out. The valve of the trap works because of fishes' behavior. Once past the inlet funnel and into the larger container or trap, a fish tends to swim close to the edge of the container. It cannot find its way out through the narrowed end of the funnel that is away from the main walls of the trap.

Still another comparison between pot and trap is worth pursuing. The fish trap is fixed in place and the media of water and fish move with respect to it. This is like the cooking pot that is also fixed in place while the medium of heat moves through it. But the cooking pot has its primary target (food) already "trapped" in it, while the targets of the fish trap start by being outside of the container.

There is still another way of summarizing this. The cooking pot container is fixed in position in a moving medium (heat flow) and its targets (food) are already present and being chemically transformed. The hot-air balloon container is moving in a relatively fixed medium (atmospheric air) and its content (hot air) is flowing in. The fish trap container is fixed in a moving medium (water) and is collecting moving targets (fish) by the actions of a one-way valve (the funnel entrance).

These variations suggest another container that is similar to the fish trap but different enough to be interesting, the butterfly net. It is a moving container in a largely fixed medium (air) with potentially moving targets (butterflies, insects). Usually, there is one opening that serves for both filling (input) and emptying (output). No lid, as such, is present. Instead, capture or containment is handled in two ways. As the net encloses, the target creature becomes enmeshed in its strands. Or the mouth of the net may be folded over, by a twist of the wrist, to seal the opening. The target is then transferred to another container, a more permanent holding jar or tank.

The handle of the butterfly net allows for the container to move more rapidly than the user. As a result, action or capture can take place at a distance from the user. For a constant rate of angular movement, the velocity of the net along an arc increases with the length of the handle. However, capture success varies in complex ways. The capture rate *increases* with the speed of the net and the number of potential targets. The greater the range of the net and the faster it moves along its arc, the more insects are at peril. The capture rate *decreases* with inaccuracy of net placement. The longer the handle, the less control in accurately placing the net because of depth perception problems and the amplification of placement error with increasing length. Thus contrary evaluative criteria—speed and accuracy—are associated with the length of the handle, but in opposite ways. The ultimate tradeoff between speed and accuracy depends on the skill of the user and the agility of the insects.

The butterfly net also has an important relative, the insect netting used for protection by people in insect-infested climates. The butterfly net starts with targets outside the net and ends with the goal of having the targets captured inside the net. Insect netting starts with bugs outside the net and the user inside; the goal is to keep the insects outside. The insect netting "contains" things by keeping them from coming in. Both the butterfly net and the insect netting serve to filter air, but for different purposes: the butterfly net to hold down air resistance and increase the net's rate of travel through the air, and the insect netting to allow the user to breath.

So far, the containers we have discussed, from cooking pots to butterfly nets, have been physical containers. They involved the capture or exclusion of matter, which may be filtered in or out, and whose passage is governed by lids or valves. Now we will take a step up in abstraction.

Recently, I read of a new method of incarceration, designed in part to cope with increasing crime rates, the overcrowding of prisons, and the high cost of building and staffing new prisons. In the new method, the convicted felon wears an electronic device on his or her ankle; the device cannot be removed without destroying it. The device responds to a particular frequency of electronic signal sent over the phone. During certain hours of the day, after work and on the weekend, the felon is "imprisoned" in his own residence. To ensure that the felon is so imprisoned, a parole officer will call at random times, electronically signal over the phone, and the felon's ankle bracelet will respond appropriately to the signal, indicating a state of confinement. Of course, if the felon has left the premises, the electronic transaction cannot be completed, and the felon may then be sent to a conventional prison where there is a more complete restriction of freedom.

How is this a state of imprisonment or confinement? Where is the container? What are the walls, where are the guards? Here the confinement is being in a specified place, at a given time, and having freedom of movement curtailed. The walls are not, however, the physical walls of the house or apartment, because they can be exited at any time. Moreover, no guard is in attendance to physically force back or shoot an escapee. Instead, the real walls and guards are a series of rules or contingencies specifying consequences of being in the right location at the right time. That these rules and contingencies are enforced is determined by monitoring electronic signals from the felon's ankle bracelet.

To see how the rules work, consider this question: How does the prisoner get out of the confinement, the container? Two ways. First, by spending enough time under the rules and contingencies—serving the sentence. Then the ankle bracelet is removed, and the person is free to come and go in the same way as other people who have served their time are free. Alternatively, the felon can "escape" by cutting the bracelet and leaping the electronically monitored symbolic walls. But if he does so, a new set of rules and contingencies come into effect, and he is likely to find himself in a more severe set of circumstances than those of the electronic confinement.

Numerous pros and cons have been suggested for the electronically monitored containment of felons, or *home imprisonment* as it is sometimes called. Arguments for it include lower costs than conventional incarceration; the diminished need for new prisons; and the avoidance of prison environments, which essentially provide postgraduate training on how to better commit crimes. Arguments against it include its obvious Big Brotherism, the danger that as it becomes more and more automated, the state will increasingly use and abuse it as a mechanism for keeping track of its people. Undoubtedly, we as citizens will want to consider some of these arguments. For the present purpose, however, electronic containment illustrates a higher level of abstraction that the concept of containment may enter into, a containment not of walls but of rules.

Some containers are purely symbolic. A mathematical set is one example. An integer either belongs to (is in) the set of even integers or the set of odd integers, but not both. Here we have defined our symbolic container by implicitly evoking a rule structure: a test that distinguishes between even and odd integers. The "container" then is the *rule of membership*. We can also construct a container by putting braces around its possibly arbitrary members: $A = \{a, f, z, r\}$ and $B = \{b, t, d\}$. The two sets, A and B, are defined by the entities that fit between symbolic walls, the braces "{" and "}" that demarcate set membership.

Other symbolic containers consist of computer data structures such as the *file* (a collection of related records, like those in a file folder), the *record* (a structure in a file, like a form for an individual), and the *field* (a structure or place within a record, like a single entry on the form). The walls that separate these nested containers from one another are simply special symbols like commas or tabs or brackets that demarcate or delimit one form of information from another.

If file structures in a database are really containers, we may expect problems similar to those we encountered with some of the more physical container forms: how to pack as much information as possible into a given amount of storage space; and how to gain access to or retrieve that information as quickly as possible. In addition, a file structure should be *portable* (usable by many programs and different computers), and its contents should be *modifiable* as new information is entered into it (new customers) or old information in it is changed (updated phone numbers or addresses). Indeed, all these requirements are paramount in the design of both ordinary storage and computer files.

We turn now to the most abstract container of all. The linguists Lakoff and Johnson mention a powerful metaphor, *the mind as a container*, and indeed the mind is filled with facts, and from time to time we need to get them out. The way facts get into the mind is through valves that we call the senses, each valve sensitive to its own form of energy: Ordinarily, the eyes do not respond to sound waves and the ears do not respond to light. The way we get facts out of the mind-container is by special utensils: the tongue for speech, the hand for writing, and the other muscles of the body for remembered actions. The subject of cognitive psychology is largely devoted to the elaboration and specification of properties for the mind-as-container metaphor.

Let's end this section with a puzzle. Look back at Figure 16.2B. It is about the size of an orange squeezer, and vaguely resembles one except that the center pole has a hole in the top, there is a lid, and the base is mounted on several short legs that allow space underneath the device. What is it? It was used frequently in the South around the turn of the century. If you cannot figure it out, see a hint in the Notes section. I'm interested in any reader ideas on the use of this curious invented form, this mystery container.

Now let's see how a very different class of entities, clothing, may be contained.

Containing and Retrieving Clothing

Numerous devices and methods are available for storing clothing. They include the simple trunk, the dresser, and the hanger-and-pole. Of the three, the trunk and the dresser are more obviously container-like, but the hanger-and-pole is a one-dimensional container system because movement along the pole is only to and fro. The comparison of trunks, dressers, and hangers will reveal a number of important principles underlying the storage, organization, and retrieval of clothing.

The most obvious container is a trunk, in basic form a large box with handles and a lid that latches. The trunk operates somewhat like the computer scientist's data structure, the push-down stack. Items are stacked or layered on top of one another during the process of storage.

During retrieval, getting the items back out of the trunk, we pop or take off the top item first, then the next item, and so on until we find what we are looking for. Stacks of this nature are sometimes referred to as Last-In-First-Out data structures. However, one important difference between the computer scientist's stack and the layered clothing in the trunk resides in the method of treating overflow. When the computer stack reaches capacity, the bottom item drops out to make room at the top for a new item. When the trunk reaches capacity, the overflow is on the top. Still the comparison is useful, because in both cases items go in from the top and are retrieved sequentially from top to bottom.

Sequential access to the items in the trunk slows down the retrieval of clothing. It also curtails the instant comparison of nonadjacent clothing items for things like color coordination. Of course, candidate pairings can be removed from the trunk and set side by side on the bed to better compare them, but this involves extra effort in removal—and also effort in returning items to the trunk's stack if a match is not satisfactory. In addition, storage in the trunk will produce wrinkled clothing due to the weight of higher layers on lower layers. In a general sense, this is a storage interaction problem. Once stored, items are not inert; they produce complex dependencies among themselves. The trunk also allows for categorical storage (all shirts together, all pants together), but it may involve difficult insertion operations to keep the members of a category together.

If we reduce the size of the trunk, copy it many times, and put the copies in an overall structure, we produce a device call the *dresser*. The dresser has several drawers, and categorical sorting (socks, skirts, shirts, . . .) is possible by drawer. Unlike the trunk's stack that must be accessed in serial order, each drawer allows random access—it is unnecessary to sort through the contents of drawer A before going to drawer B; we may go directly to drawer B. Random access greatly speeds up retrieval, if we have organized the drawers by category of clothing. Each drawer itself, however, is a small edition of the trunk or pushdown stack, and we may need to store the contents and retrieve them layer by layer within a drawer. The flexibility we gain in having a number of smaller trunks or drawers is paid for in part by sometimes having to split categories between drawers. When this becomes necessary, we have the choice of the same category being in spatially nonadjacent drawers, or of performing a major reorganization between drawers in order to have our split category occur only in spatially adjacent drawers.

Because categories of clothing may have different space requirements, the dresser often has several sizes of drawers. But any given assignment of clothing categories to drawers may need future reorganization as our wardrobe changes or as the seasons change. No provision is made for continuously adjustable category sizes. Only the existing drawer sizes are available, unless the drawers can be further subdivided with removable partitions, which may get lost or be in the way.

Since each drawer is a small stack in its own right, wrinkling of clothing remains a problem. Many of the attendant sequential storage and retrieval problems of the trunk remain, even if on a smaller scale.

Several storage problems associated with trunks and dresser drawers can be solved by placing clothing hangers on a horizontally mounted clothes pole in a closet. The pole then acts as a one-dimensional container along which hangars and clothes can move freely. On the pole, clothing can be grouped by categories simply by moving the hangers around; removal and insertion are easy operations with hangers. Pants are placed on contiguous hangers and skirts on another group of contiguous hangers. The size of a category is completely flexible (a continuously variable space up to the capacity of the pole itself) by simply sliding, adding, and removing hangers from a region of the pole. In addition, the hanger system allows for graded categories. For example, pants may be organized from cold- to warm-weather suitability. Like a library card catalog, the separate clothing items are also indexed on the pole; the visibility of a garment's part serves as the index for retrieving it. Finally, the indexing facilitates random access to an item. It is unnecessary to start at the beginning and work sequentially through the rack until the desired item is found—simply look and grab.

Because of the continuously variable category space and random access to the hangers in those categories, new clothing categories can be created at will. An example is the outfit, an ensemble of different clothing categories that go together. Another example is what-will-be-worn-tomorrow, a small collection of hangers and clothing selected the night before. In principle, night-before selection may be done with the trunk and dresser also. For the trunk, tomorrow's wardrobe is placed on the top; for the dresser, tomorrow's wear is in a special drawer. But the difficulties of sequential access for trunk and dresser make creating categories on the fly more difficult, and the likelihood of having everything wrinkled also increases. Problems of this nature mean that the user of the trunk or the dresser may well select tomorrow's wear the night before, but it will be ironed and stored *outside* its normal location of trunk or dresser.

Clearly, the hanger system is the winner over the trunk and the dresser when we need adjustable category sizes, flexible arrangement of categories, ordering within categories, flexibility in adding new categories, reduction in wrinkling, and random access. Is it all so one-sided? What advantages, if any, have the trunk and the dresser over the hanger-and-pole system?

Items of clothing like socks and underwear do not lend themselves to a hanger system. Few people worry about wearing wrinkled socks or underwear. We may have a large supply of underwear that is fairly homogeneous; hence it's unnecessary to sort through a stack to find just the right item. Then we come to socks. The idea of hanging them is strange indeed. If socks are draped over a hanger, a single pair per hanger, they will get lost by falling to the floor. A sock-pair per hanger also produces inefficient use of the space. Even if more than one pair is placed on a hanger, an inefficiency remains: Vertical space is wasted.

Socks probably work best in a drawer. Because they are small when rolled up, many pairs can be seen at once. Simultaneous visual presentation and therefore something close to random access is preserved, even if we need to occasionally stir the socks to find the right color.

Finally, if portability is important, then the trunk is the winner over dresser drawers and hangers. But what if the hanger, the dresser, and the trunk were joined together? Some of the airliner-style clothes bags now combine many of the same functions, using hangers and independently accessible storage compartments, if not drawers. A complete join of the three is sometimes called a steamer trunk or a wardrobe trunk. However, it's too big for real portability. The ultimate in this is the Lillian Russell patent for a wardrobe trunk, an elephant-sized box that unfolded into a dresser, vanity, and closet. Most likely, it was not a commercial success. Its appearance suggests that it is simply too large, heavy, and complicated for portability. It may be a join in which the worst feature of each ancestral component manages to triumph.

But let us return to hangars for a moment. They are miniature storage devices in their own right, a form of interface between pole and clothing, and they come in many varieties. The plain hanger is shown in Figure 16.3A. Its parts are: hook, neck, shoulder, and base. Usually, the plain hanger is made of soft metal wire. The plain hanger is suitable for shirts and dresses. It supports clothes with its shoulders, yet it creases pants when they are doubled over the thin base wire.

The modified plain hanger with a broad base is not illustrated. It's a simple variation on the plain hangar but its base is thicker, perhaps a cardboard tube or a wooden rod, either of which makes it possible to hang pants without wrinkling them. The principal support is now the base rather than the shoulders of the hanger.

The two-clamp hanger is shown in Figure 16.3B. A hook is attached directly to a base or to a cross-member, and from that base two spring-held clamps, one at either end of the base, hold the extreme edge of a garment. Pants and skirts may be hung without creasing with the clamp hanger.

The split-clamp hanger is not illustrated but it is a variant on the

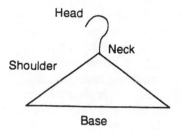

A. Plain hanger with hook, neck, shoulders, and base.

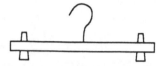

B. The two-clamp hanger.

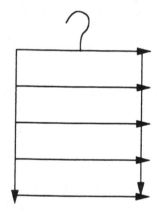

C. Multiple cross member hanger.

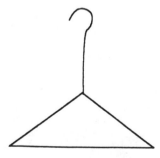

D. The long necked hanger for short people.

Figure 16.3 Variations on the hanger theme.

previous case. The base or cross member is split so the fabric is wedged between the base parts, which are then locked with a ring that is pulled down.

The suit hanger (not illustrated) consists of a hook top, usually a broad wooden shoulder, and a two-part base of wood and wire that enables one to keep suit coat and pants on the same hanger without wrinkling. The wire clamps the pants on the wooden bar to keep them from slipping off.

The multiple cross-member hanger of Figure 16.3C shows a generalization of the modified plain hanger. The base component is repeated. Several pairs of pants may be hung from a single hanger of this type. It provides a very efficient use of vertical space, but random access is impaired by the effort required to extract just the right pair of pants from one of these fiendish devices—the remaining pants may also come off or slide to one side and get wrinkled.

Consider the parts of Figure 16.3 as points of potential variation. For example, the hook is usually the interface to a horizontal pole, although some hotels, to keep people from taking hangers, employ not a hook but a small ball on the hanger. The ball fits into a clip that slides along a track. The clip cannot be removed from the track, and the rest of the hanger isn't worth taking without the clip. The base also shows a good deal of variation. Frequently it is a wire continuous with the rest of the hanger. If clothes like pants are suspended from the base, wrinkling will occur. As noted, to avoid wrinkling, the base may have a corrective cardboard tube or it may be a piece of wood instead. This suggests that the material used to make the hanger or its parts is a variable that may range from wire to wood to cardboard to plastic.

Finally, the neck has been varied in length in a child's invention that I saw somewhere. A short girl couldn't reach high enough to hang her coat on the regular clothes poles. The problem might have been solved with a stool, a door hook, or a lowered pole. Instead, the solution was to vary the length of the hanger neck by making it very long. Figure 16.3D shows the long-necked hanger, a nice invention for short people. Notice that the variation here takes place at the interface between the hangar's shoulder and hook. Often inventions focus on an interface, either within a device (the long-necked hanger) or between entities (the hanger itself as an interface between clothing and a pole). This is because an interface is a natural point of articulation, and therefore a good place to make adjustments.

Clothing organizers can tell us still more stories. These days everyone uses computers as a model for memory, and the computer is a rich model. However, its repeated use is also getting tiresome. We need the liberation provided by other metaphors. Let's see what we can do for memory by using our various systems for storing and organizing clothing. Do any of these methods—trunk, dresser drawers, or hangers-and-pole—provide a reasonable metaphor or model for human memory. After all, memory is a kind of container involving storage and retrieval of information. Is one of these methods of storing clothing a better metaphor for how human memory works than the others? Let's uncover the answer by producing questions about memory and relating our storage devices to the answers.

Is there organization in memory? Obviously, Yes. When I ask you to think about *dogs,* you probably run through a series of *related* images or associations. That is, you don't consider unrelated items like the Chinese food you ate last night or the concert you went to, both of which were between the separate experiences of walking your dog and being awoken by a cold nose on your cheek. Hence memory is not like the trunk in which everything is thrown willy-nilly and stacked one thing on top of another. It is more like the dresser or the hanger storage system where related items are in about the same location.

Is there random access in memory? The dog example is again appropriate. You didn't start at the beginning of your memory and sequentially search through it for all dog experiences. You more or less went to the relevant parts of memory concerned with dogs, as required of random access. Admittedly, we sometimes search our memories sequentially, as when we are trying to find our car keys by reconstructing our recent past. This example tells us that human memory retrieval is flexible. It can find information either through random or sequential access. In this sense it seems more like the hanger system: You can go immediately to what you want or you can carefully and sequentially compare, say, the color coordination of items.

Can categories in memory be created on the fly? Human memory is particularly good at creating new categories. An important and not completely obvious example is the category of *novelty.* Something is new to us; we recognize it as such; and we start a category of all things that are similar to the new thing. Only the hanger system allows us to create categories easily. A new category system for the trunk means that we must take things out and build our organization anew. For the dresser drawers we must have many empty drawers interspersed between filled ones or we must do extensive shifting of content and drawers to create a new category.

Does memory involve graded categories? Yes, memory is particularly good at this. The work on natural categories indicates that memory does indeed use graded categories. For North Americans, a robin is a better instance of a bird than an ostrich. Here "better" means that people are more likely to list it first when asked to write down as many birds as they can think of, or they can more quickly decide that the word *robin* refers to a bird than the word *ostrich.* Again, the hanger model is superior to the trunk and the dresser drawer models for creating graded categories, like warm to cold weather shirts.

Does memory work with partial retrieval cues? Yes, it does. The shape of a person's head from the back is often enough for us to recognize the person. Sometimes even a fragrance is enough to transport us to a richly connected set of memories. Novelists like Proust have described how the taste of a cake may evoke a stream of childhood memories. The trunk and the dresser drawers present no visible cues until we plough into them. The hanger system wins again by showing a small piece of each garment, an excellent indexing and retrieval cue.

Certainly, the hanger system and human memory share some characteristics. But is the hanger system really a good metaphor for human memory? Notably, it fails to capture the automatic nature of filing things in memory, and it does not embody the self-organizing features of memory. However, it's a good first approximation, it seems to represent many characteristics of memory, and it's far simpler than a computer model. Of course, no one is going to abandon a computer model of memory for a hanger model, because the information-processing capabilities of the computer are much more powerful. Nor has it really been my intention to bring converts to the hanger model of memory. What I have been trying to convey is that simple invention systems can offer interesting models of how our mind works. The mind produces an invention—a computer or a hanger system—and then the invention models the mind, in full circle.

The Cartridge Container

The hanger model of memory shows that inventions not only shape our world, they also shape our mental life. There is still another way in which inventions shape our mental lives. Consider the following metaphors:

> Don't go off half-cocked.
> He was only a flash in the pan.
> That program certainly misfired.
> Keep your powder dry.
> I made it, lock, stock, and barrel.
> She's a straight shooter.
> He has a hair-trigger temper.
> Chewing the rag.

These are all metaphors from an earlier day that involve firearm inventions. The meaning is usually obvious; the last one, chewing the rag, comes from wetting a cloth patch to insert down a rifle barrel for cleaning.

Some of history's best, most inventive minds have focused their inventive talents on firearms. The incentives for firearm invention are easy to imagine. The activities associated with them have always been important: hunting and food, defense and life, conquest and gain. Notable among firearm inventions is the *cartridge,* a particular kind of container that we alluded to in the discussion of the tea bag. The concept of the cartridge as a container of previously assembled material has greatly shaped the human interface to firearms—and to the interface of other consumer products as well.

Some background on the cartridge is in order, so let's start with the flintlock. The flintlock rifle is the weapon of the American Frontier. It is a muzzle loader with a flint-tipped hammer that strikes a small metal plate (the frizzen); the resulting sparks are sent into a flashpan that ultimately serves to ignite the powder behind the projectile ball. The prototype that we are talking about is a reproduction of a so-called Kentucky rifle that was built in Pennsylvania in about 1795.

In a way, a rifle is built on the idea of containment. The powder and ball of the flintlock are contained by the barrel. The exploding gases from the burning powder are contained just long enough for tremendous pressure to build up behind the ball before it is sent hurtling down the barrel. Many container-related accessories also must be available before loading can take place: a leather purselike pouch to hold implements and supplies; a powder horn to hold coarse grain powder; a powder measure; patches, or "lids," made of linen or buckskin that are soaked in tallow or beeswax for lubrication and the sealing of gases; lead balls produced by a casting mold; a flask of fine-grained rapidly burning powder for loading the flashpan; a short starter stick to get the patch and ball started down the container-barrel; a wooden shooting stick mounted under the barrel and used to push down the patch and ball farther into the container barrel; and a vent pick to clear the passageway between flashpan and the contained main powder charge.

These container-related features directly influence the human interface to a flintlock rifle. The steps in loading a ball and firing are many and complex. It's not necessary to understand each step; I simply want to overwhelm you with the complexity. Here they are, as recorded from a videotape of Keith Weaver, a restorer and builder of facsimile firearms, caught in the act of loading and firing.

- Turn the weapon on end, with the butt on the ground and the muzzle facing upward.

- Drop the shooting stick down the barrel to make sure another ball and powder charge are not already loaded—a double charge will blow up the rifle.

- Remove the shooting stick.

- Pour the powder from the powder horn into a powder measure—a small cylinder.

- Cap the powder horn to make sure that stray sparks do not ignite its contents.
- Empty the powder measure into the barrel.
- Replace the measure into the pouch.
- Put a greased patch on top of the barrel.
- Place the ball on top of the patch and over the muzzle.
- Use the short end of the short-starter stick to get the patch and ball started into the barrel. If the patch is too large, trim it with a knife.
- Use the long end of the short starter stick to drive the ball and patch a few inches into the barrel.
- Put the short-starter stick back into the pouch.
- Pick up the long shooting stick (use the foot for faster action).
- Insert the shooting stick into the muzzle and tamp down the ball and patch. Make sure there is no air gap between powder and the patch/ball, otherwise the rifle may blow up.
- Withdraw the shooting stick and replace it in the holder under the barrel.
- Pick up the rifle.
- Place the rifle on half-cock.
- Throw the frizzen forward—the frizzen is a metal plate that the flint strikes in order to produce sparks.
- Use the vent pick—a thin wire—to clear the channel between the flashpan and the main powder charge.
- Return the vent pick to its storage hole position in the stock.
- Wipe clean the frizzen with the thumb.
- Remove from the pouch the flask that contains the fine-grained and rapid burning flashpan powder.
- Load the flashpan with powder. The flashpan is under the hammer and frizzen, and it has a duct that leads directly to the main powder charge.
- Lower the frizzen.
- Cock the hammer.
- Aim.
- Pull the trigger.

The flint now strikes the frizzen, a metal plate, sending sparks into the flashpan's powder; the sparks then flash through a duct and ignite the main powder charge, firing the ball. The preceding cycle may now be repeated for the next shot. What complexity!

We tested the maximum rate of getting off three shots when going through all the recorded steps. The videotape record showed that Keith took 112.5 seconds per shot. He considered this to be about the fastest rate possible for safe firing. He told me that to go faster is to run the risk that, occasionally, sparks will remain in the barrel, and pouring new powder on top of them will lead to disaster.

Many of the steps described here can be short-circuited by a form of preplanning and assembly called *prefabrication*. In fact, the same flintlock rifle may be fired with an important prefabricated device, the primitive paper cartridge. The early cartridge is essentially a cylindrical tube of paper. It is tied just in front and just behind the ball. The area behind the ball is filled with a premeasured charge of powder, and the back end with the powder is usually just folded over. To use the cartridge, one bites off the back end, pours the powder into the barrel, and pokes the remaining part of the cartridge down the barrel with the shooting stick (the paper now functions as a patch to seal in the exploding gases). The remaining steps for dealing with the flashpan are the same as before.

Keith's time per shot, using a paper cartridge with the same rifle, was 25.6 seconds, more than four times faster. Evidently, a large part of the lengthy firing times without the cartridge was not a matter of safety so much as just needing to execute many different procedures. The cartridge, by virtue of having these procedures prefabricated, greatly speeds the rate of firing.

Notice that in all these timing trials we are talking of a *rifle*, a weapon with spiral grooves in the barrel. The *muskets* widely used in the Revolutionary War were smooth bored, that is, with no rifling. The smooth bore made the load-reload cycle somewhat faster because it was easier to push a patch and ball down a smooth bore than a rifled barrel. The tradeoff was in accuracy; the rifle gives a spin to the ball that makes it gyroscopically stable and much more accurate.

I wanted to compare the process of firing a muzzle-loading rifle with that of a breech-loading single-shot rifle that uses a metallic cartridge with a percussion cap. A typical example is the Sharps breech-loader of 1874. Ralph Shields, another local builder and restorer, showed me one in his possession. This time the load-and-fire cycle was much simpler:

- Open the breech (the end opposite the muzzle), which automatically ejects the previously fired cartridge case.
- Insert cartridge.
- Close breech.
- Cock the exposed hammer.
- Aim.

- Fire.

- Repeat cycle as quickly as needed.

Notice the contrast with the flintlock in the number of steps required! We simulated three rapid shots with the Sharps. The average time per round was less than five seconds, and Ralph protested all the time that he never practiced rapid firing. A correlated advantage of the breechloader over the muzzle loader is firing position. One loads a muzzle loader while standing, but a breechloader can be tended to while prone. In war, that means one is an exposed target when using a muzzle loader and a concealed target when using a breechloader. So many things flow from an apparently simple procedure such as loading a gun from a different end.

The flintlock rifle, even with its paper cartridge, requires many more steps than the breech-loading rifle using metallic primed cartridges. In addition to the steps eliminated by the breechloader and metallic cartridge, the switching time between operations also decreases. No longer are gross movements like turning a rifle end for end required. A measure of the time involved strongly favors the breechloader and cartridge, not only because of fewer steps, but also because the steps and the switches eliminated are among those that take the most time. Obviously, the cartridge and the breechloader are more "user friendly" than the flintlock, so the skill level and training time for the user also decreases.

Notice what the modern metallic cartridge does. It provides a standard, accurate powder measure; a built-in percussion cap; a weatherproof container; an integration of all the expendable parts of firing; and prefabrication. All these steps combine to increase reliability and user-friendliness. The metallic cartridge with built-in primer is made possible by a breech loading mechanism, and the effectiveness of breech loading is in turn made possible by the cartridge. Together, breech loading and the cartridge increase the maximum rate of the load-fire-reload cycle. They shorten the time of some operations, eliminate others, and reduce the switching time between operations.

The increased reliability and rate of firing that a perfected cartridge makes possible are paid for in part by earlier planning and activity, the *prefabrication stage*. Prefabrication combines the packaging or containment of materials and procedures in a time-shifted way. The idea of prefabrication is not original to the cartridge. Some materials and procedures were already prefabricated for the flintlock: powder was made in advance, patches were precut, and balls were cast earlier. In a later development, prefabrication reached a new height in the metallic cartridge.

During its assembly, the metallic cartridge required a percussion cap to be set, powder measured and poured, and a bullet set and crimped in the mouth of the cartridge. Undoubtedly, the manufacturing operations associated with the cartridge are faster than the comparable operations with the flintlock, because they need not be done as discrete processes with substantial dead or switching time between each step. Instead, when loading cartridges, the operations can be done as batch processes—all operations of one kind are performed, then all of another, and so on. Batch processing holds to a minimum those switching times lost in setting up different operations. However, even if the manufacturing processes associated with the cartridge were not faster, the cartridge would still have been successful because it allowed for more rapid operation *during firing,* when time was all important.

The combining or collapsing of a whole series of sequential operations into one package, such as the cartridge, is a form of *functional parallelism.* From the user's viewpoint, functional parallelism means that many things are happening at once. The act of firing is functionally equivalent to putting a primer in place (instead of loading the flashpan), measuring the powder, loading a bullet, and crimping the cartridge casing around the bullet. Even if these processes are not strictly simultaneous, all of it seemingly occurs at once when a single action, squeezing the trigger, is taken.

We can further generalize on the cartridge concept: an ammunition clip, preloaded with a number of cartridges, can be inserted all at once in some weapons. As such, it is a metacartridge or a "cartridge" of cartridges. Certainly, the cartridge clip also had to be prefabricated, a step at a time. Yet, from the user's immediate viewpoint, many cartridges are loaded at once. Another example of functional parallelism.

Can we generalize still more on the idea of the cartridge? To answer this question, we must first abstract the cartridge's essence. The cartridge concept has four essential features: first, a functional parallelism in which sequential steps and materials are combined in one package; second, the prefabrication of these steps; third, a convenient human interface to the package; and, fourth, a consumable or disposable item.

Do other inventions share these same ideas? Yes, occasionally in entirety, and often in part. The film cartridge is a striking analog to the rifle cartridge. It involves all four features. In an automatic camera the "fire and load" cycle are indistinguishable to the user. The audio cassette and the video cassette involve the first three requirements. They are not consumable, however, because they can be used over and over. (Even if people reload cartridge casings, they do not reuse the primer, the powder, or the bullet.) Perhaps the best way to express the analogy to audio and video cassettes is by paraphrasing the philosopher Wittgenstein's discussion of games: Cassettes bear a strong *family resemblance* to the firearm's cartridge. If the requirement of being consumable is completely relaxed, then the ultimate in packaging, prefabrication, and ease of use occurs in the integrated circuit, an assembly of great complexity that readily becomes a module or building block for even greater complexity.

While wondering which requirements to relax in the search for other family relations of the cartridge, I thought of still another example that satisfies all four requirements of the cartridge: the TV dinner! The first requirement of functional parallelism is evident, because all the steps that lead up to the package are transparent to the user. Certainly, there is prefabrication also; the item is conveniently "fired" by popping it into the breech of a microwave oven, and it is then consumed.

At first glance, being consumable does not seem an important characteristic of the cartridge's features. However, what the cartridge devices mentioned have in common is great commercial success, and being disposable ensures a continuing market. Much of that success is surely attributable to the efficiency of functional parallelism and the resulting simplification of the human interface, which determine what the user must learn and do. The simplification of the human interface is perhaps the most important characteristic of the cartridge concept. This simplification is possible because the cartridge acts as a container of materials and a repository of past procedures.

But is there a cost to the cartridge? What, if anything, have we lost with the cartridge embodiment? We may imagine a debate of the last century between two hunters. *Pro* admires the new cartridge for all the reasons cited here. *Con* does not like it because flexibility is lost. By loading his own rifle he can have a lesser charge for hunting squirrels than for deer. That flexibility is certainly an advantage. It is also likely that the first cartridges were unreliable. Often a new technology does not work well, but we will bet on its promise for the future rather than dismiss it because of its present performance.

Comparable tradeoffs also occur in other embodiments of the cartridge. The audio cassette, for all its portability and convenience, is very difficult to edit. Precision editing requires a reel-to-reel tape which one can lay hands on and move past the playback heads. And when a cassette jams, it is difficult to fix and usually must be thrown out. The TV dinner as cartridge also shows diminished flexibility. It must be taken all-or-none. Even if there is a component you don't like, you have to buy it anyway, if you want the rest of the package. Packaging processes or components in a cartridge often limits flexibility.

Whatever the cartridge costs in flexibility, for many people under many circumstances the benefits are worth it. When we look at the range and number of its embodiments, we must acknowledge that the cartridge concept is a commercial and human success.

It's been a long day, and it's time to relax. I've heated the water. I drop my tea bag-cartridge into the cup. Now if only there were a way of including the hot water in the tea bag.

17

Procedure's Way

I can remember a favorite TV series—Julia Child doing her culinary magic. I loved to watch her use a knife. She was a virtuoso: slice, chop, pare, almost like a ballet of hands and tools. These actions with the knife form a common, little appreciated, and powerful mode of invention, the creation of procedures.

Often it is difficult to distinguish between an invented artifact and an invented procedure. The two are intermingled, and it is difficult to know where one ends and the other begins. For example, the knife is an artifact, right? But the fact that it cuts and how it cuts depends on a number of actions or procedures—call them *slice, chop, pare, stab, bore, etch,* and *whittle.* These uses of the knife are procedural inventions that go along with the artifact invention. Likewise, the procedure of shopping for goods involves the artifact of money, which in turn is used in several other procedures, ranging from making change, to purchasing services, to gambling, to depositing in a savings account, to paying fines.

When it is difficult to disentangle invention as artifact from invention as procedure, a test is in order. For a given invention, the *Archaeologist's Test* asks: What part of an invention can be dug up 100 years from now? Assuming ideal conditions of preservation, those diggable parts we define as artifact; those that are missing from the dig we define as procedures. The Archaeologist's Test is excellent for *prospective* cases—where we have a present invention and we want to disentangle artifact and procedure by looking to the future. When we know implicitly or explicitly how something works, it's a simple matter to recover the procedures by figuring out what does not last and will not show up in the 100-year dig. However, some difficulty may occur when we apply the test to *retrospective* cases, as it does in real archaeology, because only the artifact is available—or worse, a fragment of it—and from this limited information we must make guesses about the procedures surrounding the artifact's use.

In either case, procedures are vitally important as inventions. Many fundamental characteristics of our culture are based on procedures. Procedures may be closely linked to artifact, as with those required to use a knife. Or they may be divorced from artifact, as with social convention and manners. Or they may occupy a middle ground.

Excavating a Supermarket

Let's begin by considering common procedural inventions associated with the distribution of goods, in this case, shopping in a supermarket. These are procedures of everyday life, and few if any are patentable. That does not exclude them from the category of inventions, broadly construed. Like other inventions, they are often built from simpler components, but this time the components are themselves smaller procedures. Often each individual has to reinvent them to move about in the world. We can find some good examples at the supermarket.

The American supermarket is an enchanted land for the international visitor. If you are fortunate enough to accompany such a visitor from a country without supermarkets on a first shopping trip, you will likely see wonderment: the eyes wide and searching, a smile of anticipation and excitement. Later the visitor will tell you that the variety of the market is unbelievable, the choices difficult and tantalizing.

The wonderment is justifiable and appropriate to the deep complexity of intermingled inventions of artifact and procedure afforded by the supermarket. In fact, many important procedural and artifactual inventions hide below the surface of the supermarket.

Let's look at some examples. A shopping list is both artifact and procedure. It satisfies the Archaeologist's Test for an artifact, because it can be dug up at a later time. Yet its construction and use are procedural. We must write it, and everyone knows that we spend several years of our childhood mastering the procedure of writing. When we write the list, chances are we group items together, based somewhat on their common location in the market. Later, when we use it, we do not start with the top item and search the entire market for it before going to the next item. Instead we have a procedural plan in which we start in one part of the market, move gradually down the aisles, all the while intermittently referring to our list. This allows us to minimize backtracking and unnecessary motion. However, we may make an exception by picking up fresh vegetables and frozen foods last. Also, when we have more than one brand to choose from, we have criteria for our choice, perhaps the lowest price or a favorite brand or an attractive package. When we put items in our shopping cart, we put the large hard items on the bottom and the smaller soft items on the top. When we pay our bill, we use the artifact of money or a check, but even here procedures like counting money and making change exist. When we pack our groceries, we use the bag as artifact, but the order of packing is procedural: first the big and hard items and then the soft items on top. Considerations of a balanced load also apply. We don't want all the cans in one bag.

Some of the inventions associated with shopping you may wish to take issue with. Thus you might say that shelves and aisles are not inventions. And walking up and down an aisle, while a procedure, is certainly not an invention. Even less so is the regular visual scanning of shelves. Nor is a line or a queue for checking out. I would disagree, however. Shelves are inventions; they are from the larger category furniture, certainly a human invention. As such, they do not appear in nature. They have a purpose, the storing and displaying of items in a compact and visible way. Aisles, an organized empty space, also have a purpose, providing a convenient path and traffic flow past the display shelves. To see this, consider a limiting case alternative, my favorite office filing system, the push-down stack: Simply throw everything on a pile in the middle of the supermarket and let customers sift through it as the occasion demands. Not very practical.

If your objections are not with the physical shelves and aisles, perhaps it is the procedures concerning them, walking and scanning, that bother you. How can they be inventions? In answer, we are not talking about any kind of walking here. Instead, we are speaking of that form of walking coordinated with our shopping list, that is part of a moderately systematic plan for exploring the supermarket to avoid backtracking. A limiting case alternative is to haphazardly progress through the aisles. While we may do this in an unfamiliar supermarket, we rapidly learn to be more systematic and efficient. This is also true when we scan the shelves. Once more, consider a limiting case. We do not scan every square inch of shelving as we pass. Instead, we are guided by our explicit or implicit shopping list, and we scan those areas of shelving while keeping in mind the categories of goods likely to contain the items on our list. These systematic walking and scanning procedures are all pervasive. Each individual invents them anew at an early age—and keeps reinventing them throughout life as new interests and categories form. These subtle procedures are idiosyncratic as to content but universal as to existence. Both walking and scanning with a plan in mind are complex learned procedures with purpose and action coordinated. They are procedural inventions at a deep psychological level.

Let's examine still other procedural inventions in the supermarket. David N. Perkins, a mathematician turned cognitive psychologist, asks this question: Why allow a free-access procedure to the shopper for most of the supermarket but not for the meat counter or the in-house pharmacy? Thus we move our shopping cart about at will for bread, butter, and oatmeal; but for meats, we go to the butcher counter and place our order; and for a medication, we present a prescription at the pharmacist counter. Many procedural models of shopping are possible, some appropriate for one kind of good but not another.

Perkin's question got me to thinking. An interesting exercise is to cross types of goods (regular groceries, meats, drugs, . . .) with access procedures (free access, service from a butcher-like person, or from a pharmacist like person who requires a prescription from a physician). With just the possibilities mentioned, there are nine combinations of goods and access procedures. These are indicated in Table 17.1. The combinations on the diagonal are the most common pairing of product-type and access procedure. Thus the free-access procedure is used when we buy oatmeal, the pharmacy-counter procedure for potentially dangerous medications, and the butcher-counter procedure for lamb chops. Some other pairings are occasionally used, even if they are not standard. An example is the near-prescription counter access to gasoline during World War II when one had to have rationing stamps. The inappropriate entries indicate unused or ridiculous combinations, like requiring a prescription for oatmeal. These examples tell us that a

Table 17.1   The Matching of Supermarket Access Procedures and Products

	Product		
Access Procedure	Oatmeal	Cancer Medication	Lamb Chops
Free access	*Normal	−Inappropriate	? Only if prepackaged
Pharmacy counter	−Inappropriate	*Normal	? Like rationing of gasoline during World War II
Butcher counter	Like a country store	−Inappropriate	*Normal

* = Normal mode.
− = Indicates inappropriate mode.
? = Unlikely mode.

good deal of hidden intelligence is at work in matching goods and access procedures.

The answer then to Perkin's question about why some access procedures go with certain products is a complex one. Access and product must match in terms of sanitation, safety, and expertise. Buying oatmeal is not inherently risky and it doesn't require great expertise—so allow free access. Cancer medication requires knowledge for its preparation, and application, so an expert like a pharmacist is required—control access through the pharmacist. Unwrapped lamb chops, for reasons of sanitation and refrigeration, should not have people picking them up and squeezing them like grapes, and they will also need to be weighed and priced—so limit access by having a butcher. Of course, most of these lamb chop problems are taken care of through prefabrication when they are put in an individual cellophane package, with weight and price stamped—then free access is appropriate. This last example tells us that a bit of tinkering with packaging and prefabrication sometimes enables us to move between access procedures for a given product. The movement of bulk oatmeal from a butcher-counter procedure in a country store to free access is another example.

We can readily find still more procedures at the supermarket. The checkout line, or queue, is another important procedural invention. To see its importance, examine the limiting cases. The simplest is the scramble. The first one to the cash register checks out. Another much more complex system is to have a reservation, like an airline ticket. Still another extreme is to organize the supermarket like a flea market. Each product is presided over by its own entrepreneur, who bills and collects on the spot. Intermediate forms are also possible. Thus the butcher counter often uses a place-holding system with a numbered ticket. The same system could be used for the usual checkout: Take a numbered ticket, and the checkout clerk will announce the next number to be checked through. Still other checkout procedures are possible.

Instead of having multiple lines, each going to a different cash register, a system often used by banks and post offices can be employed. Here the queue is of two stages, a long line in which everyone falls, and then a short line that has a maximum length of one person right in front of each cashier. Whenever a short line is empty, it is a signal to the next person in the long line to advance. This system has the advantage that you will not prematurely commit yourself to a line that turns out to have a price-check or a check-cashing problem. It also has the advantage of parallelism at the bottleneck stage of being tallied at the cash register. Still there are reasons for thinking that the usual check-out procedure with multiple lines and multiple registers might be the best method for groceries and carts, particularly if space is a problem.

The number of procedural inventions for queuing is very large. For some situations, we find some queuing methods better than others. Like artifacts, procedures become specialized and integrated with one another in endless ways.

To drive home this idea, one more example of procedural invention is in order. Let us compare being served a *prepared* meal at a restaurant with picking up *unprepared* food in a shopping basket at the supermarket. We ask a familiar question: Is there some procedure that is somewhere between the two, a procedural interpolation? Yes, one example is the cafeteria. There we serve ourselves already prepared food. So the cafeteria seems to be a partial integration of the other two methods of getting food. This suggests another question: Is there an in-between procedure with unprepared food? Yes, it has us sitting down, like at a restaurant, and filling out a slip on which we order the wanted items. Then a "waiter" for unprepared food brings us our bagged order. Interesting. It looks like procedures can be invented by combining elements, perhaps through joining and blending existing procedures, just like the physical inventions we have already studied.

The implicit procedures associated with the supermarket become more obvious when we once more apply the Archaeologist's Test. A volcano, called Vesuvius Here, has erupted. The supermarket has been covered with ash. One hundred years later, a team of alien archaeologists unfamiliar with supermarkets has begun excavation. They are finding many metal and glass containers; these are clearly artifacts. The team is unfamiliar with the supermarket, because they do their shopping by computer. For this reason, they do not know about any of the procedural inventions we have presented. They must try to reconstruct retrospectively the procedures from the artifacts and their fragments. After much effort, they arrive at their conclusions. They have succeeded in excavating an important religious shrine. The wire baskets were used by people to bring offerings to their gods. Those offerings were in the form of food gifts brought in carts by acolytes to the shrine. After stacking their gifts on shelves especially constructed to receive the offerings to the gods, the people would kneel in front of the cash registers where the priests handed out round silver icons that commemorated dead kings, and a very occasional queen. The alien archaeologists are happy. From the artifacts they have successfully reconstructed the procedural lives and inventions of this strange people.

From Handcraft to Assembly Line

As the story of the supermarket reveals, invention is more than artifacts. Instead of the procedures concerned with the distribution of goods, we now turn to procedures associated with the origin of goods, that is, with assembly.

To me, what is most striking about a modern manufacturing plant is not the sophisticated machinery at work, but the coordination among people and machines. In this respect, perhaps no manufacturing process so much invisibly shapes our lives as the *assembly line*, a way of manufacturing that produces abundance at low consumer cost. Unfortunately, it sometimes produces as a by-product a mind-numbing automatism in the workers that feed its appetite.

We tend to think of the assembly line as highly automated, a succession of robots working on an automobile body as it moves continuously down a conveyor belt. But no necessary connection exists between the assembly line and either mechanization or automation. The historian Siegfried Giedion cites as the assembly line's origin the slaughter houses of Cincinnati in the 1830s. "After catching, killing, scalding, and scraping, the hogs are hung from the overhead rail at twenty-four-inch intervals and *moved continuously past a series of workers. Each man performs a single operation*" (italics added). The first worker splits the animal with a knife, the second removes the entrails, the third takes out the liver and heart, the fourth hoses out the body cavity, and so on until the parts are as small as desired.

Strictly speaking, this is not an assembly line but a disassembly line. From the hog's view, the distinction is important; for the principles we want to discuss, it does not matter. The critical points here were italicized: the work moves continuously past a relatively stationery worker, and each worker performs a single operation. Contrast the assembly-line method with the other manufacturing methods discussed earlier. A discrete processing mode involves killing and butchering one hog at a time; in the limiting case, a single worker performs all the operations. In contrast, a simple batch processing mode has one person killing a series of animals in place, then performing the next operation on each animal in turn, and so on for every successive operation. The transition to the continuous assembly line is not dramatic, but it is radical in result. From the individual worker's view, it is a batch process in the extreme: he does one operation, and that is all that he does. But something more is involved. The worker doesn't have to move around the manufactured good because it comes to him or her. From the view of the moving assembly line, the process is continuous, and the work never stops while the line is moving. From the same view, the assembly-line process involves a large amount of parallelism. Many hogs are in circuit at any given time, and many agents or workers are operating on the hogs at once.

Nothing is highly mechanized here. The critical operations are still done by the human hand. So what is the advantage of the assembly line for slaughtering over a conventional method of slaughter? Exactly the same as that for the eyed needle and continuous sewing over tying individual stitches: the elimination of switching motions and time at several levels. First, because the line moves the hog to the worker, rather than the worker moving around hog, the worker's motions and the switching time between motions are minimized. The proportion of on-task activity is greatly increased, although it is a matter of degree. Second, because the worker performs only one operation, mental processes do not need to be switched from one task to another. Research has shown the considerable costs of mentally switching between different thought processes, as compared to the savings that accrue from staying in the same mental mode.

Another savings that results from doing only one operation is the specialization of labor, dividing up a task among different workers. Specialization on simple tasks allows for the development of expertise and speed, with a minimum of training. Of course, to develop the concept of one-operation-at-a-time requires the ability to parse or break down successfully a complex activity into meaningful components. Each component must take about the same amount of time to do, because each worker should finish the allotted task at the same time. Otherwise, the assembly line will have to be slowed to take account of the operation with the longest duration. The assembly line, then, is also based on organization and coordination.

Still there is more to the assembly line than efficiency. My father worked in the auto plants of Detroit during the 1920s and 1930s, and I remember hearing some of his stories about the downside of the assembly line. Drawbacks included an ever-present tendency of plant managers to speed up the line. This took mind-numbing work to a breakneck pace. It is one of the factors that eventually led to unionization of the auto plants. The tendency for managers to speed up the line was natural. As workers became more expert, they were capable of working faster. If they could work faster, it meant more output and profit from a given facility. If a worker could not keep up, well, there were always others to replace him.

Automotive assembly was a much more sophisticated procedure than that of the earlier packing plants. An early setup at an engine block assembly worked this way. A main line on which the engine block was assembled consisted of a huge conveyor belt. To the side of the belt were overhead tracks or chutes into which individual parts were fed to a work station. A worker would grab a part and walk along the main line, securing the part with a bolt and nut to the evolving engine block, and then rapidly walk back to his station, to repeat the process. Notice that this is not quite a stationery worker with work moving past him; he must march with the line. A more efficient procedure, which followed later, had a gate come down in front of the object of assembly to keep it from moving forward on the line while it was being worked on, even though the line itself still moved continuously. After fastening the part, pressure on a foot peddle lifted the gate.

This arrangement permitted less worker movement and more relief from the dead time of walking back to the work station, although the overall work was strictly paced by the continuously moving line. There was parallelism here also. The engine block was worked on by several people at once at a given station. The person parallelism was limited by the size of the engine block and that in turn limited the number of agents working on it.

Parallelism was especially prevalent for chassis assembly. A number of workers at a station performed different operations simultaneously on the same chassis, and from different sides of the line, because the chassis' large scale allowed for multiple workers engaged in simultaneous action.

The burnout rate was considerable. Not everyone could keep up with the assembly line's speed. A worker's failure could be determined quickly, if it were a physical limitation. But boredom coated with stress might take longer to develop. However, the pay was good in comparison to other jobs, so people did their best to stay.

One of the mental difficulties encountered was undoubtedly an incompatibility between the nature of the human attentional system and the continuous unrelenting movement of the line. Human attention fluctuates in ways that are not completely under our control. Sometimes we are on task, and other times our attention wanders. Ideally, we want to control the flow of our work to match the highs and lows of our attention cycles. We can do that with the discrete assembly of handcrafted items, but it is not possible with the assembly line where the actions of many workers must be coordinated in time. To produce a successful product, one without faults, human attention must march in cadence with the moving assembly line. The difference between the voluntary control of attention in handcraft work and its compulsory driven quality on the assembly line is probably one factor that makes discrete production of handcrafted items more psychologically satisfying and far less stressful.

Now let's generalize on the idea of the assembly line. One of the main ideas is coordination: Everything must arrive at just the right time for the agent at a station to perform the right operation. The idea can be extended. The assembly line, as concept, does not need to be confined to the walls of a building. The sources and influx of raw materials and already assembled subcomponents form part of a broader line extended in space and time beyond the walls of the factory. This broader line is in fact one of the key ingredients of the Japanese system of manufacture known as *just-in-time supply*. Just-in-time allows the manufacturer to hold inventories to a bare minimum, holding down storage costs and billing costs for parts not yet needed. But it requires a coordinated network of suppliers who can deliver on time. If a critical component is missing, everything may have to grind to a halt. The planning and coordination required of this system means that the flow of control outside the assembly plant must be as carefully thought out as it is inside the plant and on the assembly line.

Hence just-in-time inventory can be thought of as a simple generalization of the assembly line concept beyond the confines of a factory. The generalization works in another respect also. Some of the assembly line's problems also occur for the just-in-time supplier who must accumulate inventories to avoid delays for the main manufacturer.

For generalization, however, the easiest place to start is the type of product assembled or disassembled. Any complex artifact that is made of many components, or that requires many operations on it, is a candidate for an assembly line. The fewer the components or operations, the less the benefit from an assembly line. Further, the more items to be produced, the more worthwhile it is to go to the trouble of setting up a line. If only one or a few items of a kind are to be produced, there is no point to investing in the setup time.

Maintenance to Extend the Life of an Artifact

We have talked of the ideas that lead to procedures, of the procedures used to distribute and assemble goods and artifacts, and now we need to talk of another important procedural idea: how to extend the life span of the artifact we have invented, manufactured, and distributed. When did the *idea of maintenance* begin, and why?

Tools used by animals do not need maintenance. The tools of chimps are made and used on the fly. A leaf is chewed to make a porous sponge, a branch is broken from a tree to use as a club—and after use the tool is discarded. For the chimp, a tool is not an enduring artifact. The earliest stone tools of our hominid ancestors also indicate on-the-spot manufacture at a use-site and then abandonment. The evidence for this claim comes from the finding of manufacturing chips on site and little usage as revealed by microwear analysis. But later, tools were made ahead of use as shown by many chips at a central manufacturing site. Often that site was separated by miles from the source of the stone material. The finished tool then had to be carried to the location of use. In addition, signs of persistent use of a tool have been revealed by microwear analysis of the stone. Together, these findings indicate an enduring artifact that required foresight, skill, and effort invested in its manufacture. Such a tool is valuable and needs to be preserved. It is far removed from the ephemeral tool used by animals.

Not surprisingly, the idea of maintenance is an old one, with some Stone Age tools showing signs of maintenance. For example, a broken blade edge is retouched. Or a 20,000-year-old ivory needle shows evidence of a broken eye and later reboring for a new eye.

So why does maintenance occur? The rational answer is when the cost in effort or resources is greater for replacement than for repair. If maintenance is to be employed, the idea of the human interface must expand to incorporate an invisible subcategory, ease of maintenance. Sharpening a stone knife by knapping or pressure flaking requires much more skill than grinding does. Almost anyone can perform a grinding operation. Grinding must be a safer operation too, with a lower probability of breaking the tool than for an operation like re-chipping. When maintenance is difficult and an item is expensive, whole occupations are created: possibly flint knappers and certainly auto mechanics. The difficulties of maintenance are starkly revealed in complex artifacts. If cars were not so expensive, people would gladly throw them away when they break down, much as they did an earlier broken arrowhead.

Somehow, we must deal with the issue of maintaining our artifacts, and people have been remarkably inventive in their approach to this problem. Consider some extremes. If ease of maintenance is taken to the limit, the result is a maintenance-free tool. Maintenance-free tools come in two very different forms, *throwaways* and *last-forevers*. Cheap forms of mechanical pencils, and even tape recorders and watches, are now expendables, either because they are designed that way (pencils) or because they will cost more to repair than to replace (low-end watches and tape recorders). As to design, making a mechanical pencil that allows for maintenance is likely to be more expensive than making a throwaway.

At the other extreme is the last-forever. We can now create expensive items or processes that are "maintenance-free" and are not throwaways. The coating of synthetic diamond film on tool edges is an example. The process is now employed on expensive items, like oil-drilling bits, but we can expect the cost to go down and the process to extend to more widely used artifacts. Imagine a kitchen knife with such a coating. A knife with a diamond-coated blade will be virtually maintenance-free. It will never have to be sharpened, or at least not for a very long time. There is sure to be some manufacturer resistance, however. A diamond-coated razor blade might last for the life of the user, but the economic underpinning of razor-blade manufacturers is based on blades that rapidly rust, wear down, are thrown away, and then replaced. A lifetime razor blade may be a manufacturer's worst fantasy.

But let us return to the more common reality of maintenance, a middle-edged reality that involves neither the wear-ever nor the throwaway product. What can be done to improve the maintenance of these products? Easy maintenance procedures can be designed into the product, but they rarely are. Consider first an example of bad design for maintenance and then an example of good design. My car, an '87

Nova, has a dipstick problem. The handle end of it is down low and hard to see; the place that the stick inserts into the engine block is even farther down and harder to see. Even if I can see the handle of the dipstick to grab it, I still must reach through a small jungle of wires to get it out—and use the same routine to get it back in. It is a bad maintenance design; surely there is a better way. A deeper question: Why should a problem like this occur at all? There is nothing new about dipsticks.

By now, we need a contrasting example of good design for maintenance. A common source of exasperation for me is losing a button on a shirt. I can sew a button on, but I rarely have one that is the right color or size. Recently I bought some shirts with two extra buttons sewn on the front tail. What a thoughtful idea! A built-in supply of spare parts to simplify the maintenance procedure.

Let's play with this idea of attached spare parts. Why buttons? Why not other shirt parts like collar stays or extra cuffs to repair the frayed ones? The answer is an interesting one, and it is not simple. A collar stay is not readily stored on any part of the shirt, other than in the collar. Extras can be included in the package, but this is not as convenient. And collar stays are not lost that often, so why bother? An extra cuff is not practical either. Where would we store it? Moreover, two cuffs will be required, one for the left arm and the other for the right. The cost is going up. Even if an extra cuff is available, the skill and equipment needed for attaching it is prohibitive for many users. All told, a bad idea.

The extra buttons on my shirt pose none of these problems. They are readily stored on the shirt in an unobtrusive place, and they are low in cost. They can be attached with a minimum of skill and equipment. As important, the failure rate for buttons is higher than for other parts of a shirt. All of these reasons converge to make the procedure of adding spare buttons an act of intelligent maintenance.

We can generalize more. What other products should come with added parts for maintenance? Certainly, the spare tire is related to the spare button. It is a relatively inexpensive item compared to the overall price of a vehicle; tires fail independently of other components; and they fail fairly often. For all of these reasons, the spare tire is a good idea.

But what about spare resistors, integrated circuits, or power supplies with the new TV or the new computer? None of these really seems like a good idea. The reason can be extracted from the shirt as a model case. The failure rate of the different components in a TV or a computer probably is not as skewed as that for the buttons on a shirt, so it's not clear what spare parts should be included as extras. To include a variety of expensive spare parts will not be cost effective. In addition, the complexity of replacing a defective part is great. It frequently requires special testing equipment and skills, unlike the shirt buttons.

These examples reveal that the shirt-button model of maintenance is not generally exportable, however elegant it may be in its own context. So we are driven to extremes, like having throwaway plastic knives or diamond-coated tool blades that will not require maintenance at all, or other people to do the maintenance on our goods. Too bad. Nonetheless, consideration of the extra buttons and the comparison with the idea of spare tires led me to another thought. Before beginning to write about maintenance, what would I have said is common to buttons and tires? Probably something superficial like roundness. If pressed, I would have tried to connect them at the level of function: Their roundness enables them to move with a minimum of friction. The tire rotates smoothly over a road; the button, while it does not rotate, moves smoothly through button holes because it has no corners. If pressed further for commonalities, I may have given up. Almost certainly, I would not have connected buttons and tires through their respective maintenance roles: Often buttons come as attached spare parts, and almost always cars come with a spare tire. The two, buttons and tires, are linked by their *spareness*. We need more inventions that embody spareness.

The next set of maintenance procedures incorporates the ideas of self-testing and remote testing. As invented things become more complex, the need for special testing equipment to maintain them also increases. I cannot begin to repair my TV. Even if I knew enough about electronics, I would need special equipment to test its circuits—and then perhaps some special tools to do the repair. In contrast, I can almost repair my computer, which is at least as complex as my TV.

How can that be? Through the miracle procedure of self-testing. Some computers, when they first start up, will do a self-test of their memory circuits. If the test indicates proper functioning, the start up continues uninterrupted. However, if a memory chip has failed, then the user will receive a message to that effect. Very convenient, but it could be better. The failure message is often in the form of an obscure hexidecimal coding of the failed memory address, while what I really want to know is which memory chip needs replacement. How much more convenient if a rough map of the memory circuit board appeared on the computer screen and the offending chip's location flashed on and off. Certainly, this is within the realm of technology. It is a *self-test*, and more inventions need to have built-in-self-testing.

But making things user serviceable has its limits. Manufacturers don't want people poking into the innards of TVs because dangerous voltages hide inside. Still, even qualified service people can profit from better self-testing. As the cost of labor climbs, why should the technician have to look up something in a manual when the same information or better can be flashed right on the display screen of the malfunctioning device? The self-test procedure is a very powerful idea and needs to be extended to more things, beginning with the most cost-intensive parts of maintenance.

But some devices cannot be taken to a repair shop. The cost of technician travel to a repair site is expensive, particularly if the needed part is not brought the first time. Large copying machines are notorious for this kind of problem. The technician comes on site and then finds that the needed part is back at the shop.

We need a better procedure. It is a useful relative of the self-test and is called the *remote test*. The copy machine, when it malfunctions or when it is due for routine checking, should have a special connector to a phone line. A number of test signals will then be sent to the machine for processing. The machine sends back over the phone the processed signals for evaluation. If something is wrong, then the technician has an idea of what parts to bring to the repair site. Ideally, another procedure will build on top of the remote test: the results of the phone check go immediately into the manufacturer's database, so a record of faulty design or manufacture can enter directly into the quality-control process. Many problems could then be corrected during manufacture. When it comes time to produce a new model, one of the first procedures would be to check the maintenance reports of the existing model.

An interesting variation on the remote test is now performed on the body. It is an electrocardiogram in which the signal is sent over a phone line for sophisticated computer analysis. This is a clever idea that should hold down costs by avoiding the need for expensive signal-processing equipment in each doctor's office—all of which leads us to a discussion of criteria.

Underlying our discussion are implicit criteria for a maintenance profile. The criteria should minimally include ease of maintenance (perhaps a rating), cost of maintenance ($), cost of replacement ($), probability of failure, mean time between failures (duration), self-testing (yes, no, degree), and remote testing (yes, no, degree). By assessing these criteria for a given invention, one can arrive at a maintenance profile and determine possible directions for improvement.

The usefulness of constructing a maintenance profile is shown by asking why stereo receivers, until recently, rarely incorporated tape players. A plausible answer is that tape players are largely a mechanical system, and mechanical systems typically show a poorer maintenance profile than electronic ones. Because of that, when a mechanical component fails, it drags down with it the rest of an otherwise reliable system. There is a reason why we have not mixed tape decks and receivers. The moral is to keep components with very different reliabilities away from one another.

We can now draw a parallel between maintenance procedures and

serving procedures in the supermarket, where we crossed serving procedures with product types. Not every maintenance procedure goes with
every product or service. Spare buttons attached to a shirt constitute a
good idea, but spare parts attached to a TV are not sensible. In contrast, remote testing for a complex piece of equipment is a good idea,
but we usually don't want a decision on shirt buttons determined by a
remote computer.

The mapping between artifact and maintenance procedure is subtle, just as that between serving procedure and product in the supermarket. Undoubtedly, many more general principles of procedural invention await analysis and extraction.

18

Transgenic Myth to Transgenic Mouse

The last few years I've been watching with ever more interest the research on genetic engineering that produces *transgenic life-forms*. Transgenic forms are the fascinating creations that result from mixing genetic material between different species. Stanley Cohen and Herbert Boyer are usually given credit for starting it all by transplanting genetic material from a toad to a bacterium, in the 1970s. No, the bacterium did not look like a tiny toad, but it did incorporate the toad's DNA into its own genes.

My favorite example of this audacious work comes from the first-ever patent for higher life-forms, one issued to Philip Leder and Timothy Stewart in 1988 for what is now called the Harvard Mouse. Essentially, it is a genetically engineered mouse that has incorporated oncogenes (cancer-predisposing genes) taken from the cells of a human cancer patient. (Related methods were used to produce the results shown in Figure 18.1, a mouse with a huge breast tumor that has its origin in the human genetic material incorporated into the mouse.) The Leder and Stewart patent describes a method for taking a gene extracted from humans, perhaps from people with a particular kind of cancer, and then manipulating and transferring the gene to a fertilized mouse egg. Later, a significant number of the genetically engineered mice show the human gene in their own germ cells and body cells. Those mice were particularly prone to develop a form of breast cancer that does not occur in mice lacking the human gene. The gene is not

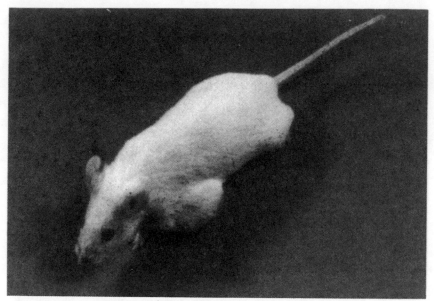

Figure 18.1 The Harvard Mouse with a large breast tumor promoted by a human oncogene. (From J. Marx, 1984. *Science, 226,* 823; by permission.)

the whole story, however; it seems to be a necessary but not sufficient condition for the breast cancer because not all of the mice with the gene come down with the cancer.

More of this later. For now the question is simple enough: Where did this idea of combining human and mouse genetic material come from? This is another story with recent headlines and an old history formed from many threads.

Transgenic Myth

While the methods of genetic engineering are new, the idea of combining the characteristics of different creatures is an old one based on the creation of mythic forms. Figure 18.2 shows what may be the oldest known transgenic creature, the Sorcerer of Trois-Freres, a transgenic form from the Stone Age found on a cave wall in France. It has a stag's horns, an owl's face, a wolf's ears, a bear's forelegs, a horse's tail, and a human's legs and feet. In our terms, he is a six-way join of these creatures. In addition, the Sorcerer is a join of procedures in his construction, a combined engraving and painting. First, he was engraved in rock; then the engraving marks were filled in with a heated grease paint.

Another example of a transgenic form is the Assyrian Winged Bull shown in Figure 18.3. He dates much later to about 800 B.C. and has

Figure 18.2 The Sorcerer, quite possibly the earliest transgenic form. (From L. S. B. Leakey. Singer et al., 1954; by permission.)

the head of a man, the wings of an eagle, the body of a lion, and the legs and feet of a bull. In our terms he is a four-way join. Why this particular combination? After all, if we use four animals and take one part from each, there are $4 \times 3 \times 2 \times 1 = 24$ possible combinations. What's so special about this one? To provide an answer, I've generated some other possible mythic forms from that set of 24. Figure 18.4 shows a figure with the head of an eagle, the arms of a man, the body of a lion, and the legs of a bull. And Figure 18.5 depicts another variation, with

Figure 18.3	The Assyrian Winged Bull: the head of a man, the wings of an eagle, the body of a lion, and the feet and legs of a bull. (Drawing by S. Manley.)

the head of a bull, the arms of a man, the body of an eagle, and the feet of a lion. The original of the Assyrians is a work of grandeur and awe; my efforts are comical.

Did the Assyrians consider all twenty-four possibilities and select the best one? I doubt it. To cast this problem as a set of twenty-four possible creatures is a modern way of looking at invention, through the idea of an overall problem space of possible forms. Most likely, their process was more like this. The king must be immortalized in this work. What is his most distinctive feature? His head. What does the king want to do or be like? He wants to fly, so an eagle's wings should be included. Also, the king is strong and powerful, so animals like lions and bulls should be included. But which should go where? The feet of a bull are more distinctive than those of a lion, so the feet and legs will be of a bull, and the lion's powerful body will take the last

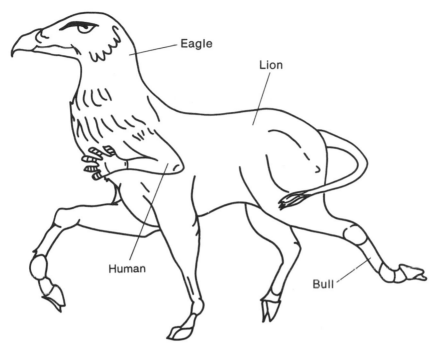

Figure 18.4 A variation of the winged bull, with an eagle's head. (Drawing by S. Palmer.)

position. Notice a familiar invention process at work here, combining parts to satisfy evaluative criteria. Only this time the criteria are a bit different: pleasing the king and producing an awesome creature. My forms are more suitable if the evaluative criteria include comic relief. The Assyrians knew what they were doing, even if they did not consider all twenty-four possibilities.

While the precise functions of the Sorcerer and the Winged Bull are not known to us, they clearly illustrate the transgenic idea. Of course, it is an idea not yet embodied in a living organism but instead as a drawing or a sculpture—that is, as a concept or an appearance of what might be. Surely, a good beginning point. But does it really have any importance to rival that of modern-day genetic engineering? I believe so. These mythic forms are an important part in the development of art and religion, two human concerns of inestimable importance. And I believe these abilities to abstract properties of forms and combine them in new ways to produce a Sorcerer or a Winged Bull are part of the same thinking processes that lead to the mouse with a human oncogene.

To see this whole fabric, let's take up another thread in the invention of transgenic forms, one we may describe as the domestication of symbolic life forms. A useful goal is to construct or invent copies of

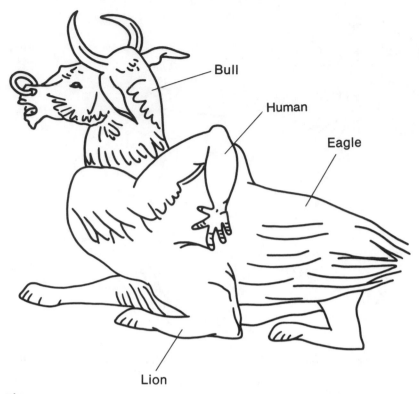

Bull

Human

Eagle

Lion

Figure 18.5   A second variation of the winged bull, this time with a bull's head. (Drawing by S. Palmer.)

the outward shape of another organism. Not only is this done for mythic purposes but for practical ones as well. A notable example is the *decoy*. Decoys are not new in human history. The early Egyptians used them for the capture of waterfowl with nets. And decoys may well have been used in even earlier times. Decoys function as *attractors*. Some resemble food and others do not. An example of resemblance is an artificial lure on a fish hook; it looks like, say, a minnow. But food is not the sole basis of attraction. The Trojan Horse excited curiosity, and it was undoubtedly of greater interest than a real horse. The generation and manipulation of form can have many uses.

Another type of attractor resembles the animal to be attracted. Here the prototype is the wooden duck decoy. Notice the complexity surrounding this invention. For it to work, there must be prior knowledge of the behavior of migratory birds. They are social creatures and congregate with their own kind. When their kind are present, it is safe for birds on the wing to fly down and join them. Also, there must be in place efficient technologies to capture or kill the birds once they descend. These are the technologies of nets and bow-and-arrow. Related behavioral and biological developments also occur in the training and breeding of special dogs to retrieve killed or wounded ducks. The decoy is but one part of a very complex system of artifact and procedural invention.

We have seen that decoys attract. Following our usual heuristic procedure, we may ask, do decoys have an opposite? Yes, objects that repulse. An anti-attractor, or a *scarecrow*, is in a sense a reverse decoy. In many parts of the world scarecrows are still frequently used to keep birds from a crop. Yet smart birds like crows learn to discriminate between real people and scarecrows, evidently by movement. To counter their learning, the scarecrow must be moved every day. Obviously, such movement is effortful, something to avoid.

A design is needed to take advantage of the wind so breezes can automatically and randomly move the scarecrow. Often this is done by having loose cloth on the scarecrow, cloth that flaps in the wind. Still, the crows are not deceived for long. More realistic humanlike movement is needed. Perhaps putting wind vanes on the head of a scarecrow, like a windmill, will serve to rotate the form as well as make it flap? Unfortunately, some crows will now learn that it is safe to land when a moving form has a vane on the back of its head. To counter this tendency, Chinese farmers often wear hats with vanes or eyes on the back. They then resemble their moving scarecrows and lend them authenticity. A complete cycle has been traveled—the starting point at which scarecrows resembled people to people now resembling scarecrows.

Scarecrows are not the only reverse decoys. Stuffed owls are also used to drive off birds, and gardening catalogs now frequently offer very realistic rubber snakes for the same purpose. Would the snakes need to be moved intermittently to keep ahead of the learning curve of birds? Probably for some birds at least. If birds did not learn of the fakery involved, it would be a simple matter to keep pigeons off civic buildings by putting up a few rubber snakes. Instead, elaborate spikes and the like are needed. This last example also shows a slide in the invention framework for reverse decoys: The purpose of the reverse decoy glides almost imperceptibly from protecting food crops to protecting buildings and walkways for health and aesthetic reasons.

The decoys and scarecrows discussed are all based on constructing a visual appearance, but these ideas may be extended beyond a visual mode. In fact, some decoys are based on generating sound. The *sound decoy* is exemplified by duck calls, elk calls, turkey calls, and squirrel calls. The reverse forms, *sound scarecrows*, are used for moles and birds. My town recently installed a siren system to keep starlings from nesting in the downtown business district. It didn't drive away the starlings but it did keep me away.

Still other senses come into play. We deploy *scent decoys* to attract deer, beaver, fish, and, if you believe in the power of perfumes and aftershave lotions, we can point to a staggering array of human attractants. *Scent scarecrows* work too. My wife uses mothballs to disrupt the olfactory system of armadillos, keeping them from digging for grubs among her chili plants.

Domestication, Large and Small

Even though invented decoys and scarecrows are pale copies of flesh and blood, they are quite useful. However, the goal of shaping or changing the characteristics of a species certainly involves more than drawing or carving its outward form. The next step up in the shaping or inventing of life forms goes beyond facsimile copies and entails the manipulation of actual living organisms. Domestication began about ten thousand years ago, with the dog. The usual biological evidence for domestication is a systematic departure in skeletal features from those of an allied wild species.

Let us start with a definition. In a strong sense, *domestication* means having human control over the breeding, feeding, and ranging of animals, so we may shape them and use them and their products for our purposes. A similar definition applies to plants. Probably, this whole process started in a nondeliberate, nonconscious way, beginning with the early images and carvings of prehistoric caves and the later use of totems and decoys. Then one can imagine collecting pets, perhaps relatively docile canines or baby animals that were fed and stayed in the area as a result. Those animals that we have selected to feed then will be in proximity for breeding. Control of range is accomplished by feeding and penning, allowing ready access to the animals instead of having to hunt them over a wide range. What may have begun for one purpose, the keeping of pets, then slides into another—that of domestication.

Of course, we domesticate for a purpose, to obtain products from the animal. The products can range from meat and fur, a one-time and lethal use, to the harvest of replenishable resources such as milk and wool, to behaviors associated with warning or sentry duty, protection and companionship, and finally to the biochemical products from a genetically engineered species.

The cornerstone of domestication is *selective breeding*, in which some fixed or constant aspect of an animal comes to be regarded as a variable subject to human modification. This insight—recasting fixed features as potential targets of variation—usually seems obvious after the fact. But before-the-fact it is anything but obvious.

Take the domestication of cattle and goats. Wild horned animals are dangerous. The Texas longhorn has magnificent sharp horns that are capable of lethal damage, and its prehistoric forbearers must have been even more intimidating. The breeding of short-horned cattle may have been initially accidental but soon became purposive.

In fact, some of the best evidence for early domestication comes from the archaeological finding that cattle and goat horns became dramatically shorter in a brief period of history. Something more than natural selection was at work, because very short horns in wild individuals must be negatively related to natural selection. And something more was involved than humans simply associating with the more docile and shorter-horned specimens. Undoubtedly, the additional factor was selective breeding for safer horns, and perhaps for a milder disposition. Both of these criteria, short horns and a mild disposition, promote the survival of humans rather than the survival of the animal in its natural environment.

But before *systematic* selective breeding can take place—this is the important point for invention—it must be realized that horn length is not a fixed or uncontrollable aspect of nature. It is a variable facet that can be tinkered with by selecting parents for shorter horns and selecting their offspring for the same characteristic. Putting these notions together suggests a domestication heuristic: In an otherwise desirable creature, notice that any undesirable "constant" feature, like long sharp horns, actually varies about some central undesired value (the average length and sharpness of horns). Try to control or shape that variation in a desired direction (shorter and less sharp horns) by selective breeding.

Human selectivity then replaces the natural selection of evolution. Whether the selectivity begins unconciously or not, it soon becomes a deliberate strategy. The naturally occurring variability of horn length is controlled or filtered to produce the desired outcome of short, dull horns, or no horns at all. The same principles also apply to naturally varying behavioral characteristics, like a docile disposition.

Domestication through selection may be considered as an important special case of variance control, a principle discussed earlier. For a variable like horn length, we shape the distribution of lengths to our needs by controlling the breeding of our stock. This is true not only for quantitative variables like horn length; variability in quality is also more easily controlled in domestic animals than wild ones.

To return to the purposes behind domestication, probably cattle and goats were domesticated for meat and hides, and later for milk and wool, replenishable products. Other by-products may also be replenishable; an example is manure, usable as a fuel for fire and as a nutrient for plant growth. Behavior is another another class of products, as evidenced by the case of the dog. While the dog was undoubtedly eaten on occasion, its most desirable product may have been associated with behavior—a sentinel, a defender, or a partner on the hunt. These are replenishable products that may be "harvested" many times.

So far we have considered common middle-sized animals as producers of either replenishable or one-time products. What other life-forms can be domesticated, and what other products can be produced? Let's generalize a bit and pick some important and little appreciated instances of domestication. Perhaps something that is very small and has some of the same behaviors as the dog, particularly the guarding function?

The mandarin orange's sweetness acts like a beacon to many pests. It is a valued crop and finding a way to control its pests is important. In fact, for millennia the Chinese have used the citrus killer-ant to protect their mandarin orange trees from other insects. The citrus killer-ant is carnivorous; it attacks and eats other insects. To facilitate the behaviors of the citrus killer-ant and help keep it in a given area, long bamboo poles are placed between trees to provide paths for the ants to move quickly from tree to tree so they may prey on invading insects. Evidently, the ants fulfill their function admirably and have been used as biological pest controls for centuries. Here the product is again a behavior, the insect-hunting and eating behavior of the ant. I'm not sure that the Chinese control the ant's breeding. If not, at a minimum, they trap them and place them in a desired environment. If breeding is not controlled, then the citrus killer-ant enjoys a weaker form of domestication than its cousin-by-behavior, the dog. By relaxing the requirement for breeding control, the framework for domestication has slid into something a bit different, but we know that frequently happens with invention.

Another example from the Chinese, with full domestication afforded by breeding control, is the silkworm, the larval stage of a moth. The silkworm spins a very fine thread of great strength to form a cocoon. The thread can be unwound from the cocoon by soaking it in hot water. It is then spun together with other threads to form a thicker thread, and ultimately these composite threads may be woven into a fabric. The silkworm's thread serves as a component or building block for the assembly of a more complex material, silk cloth. In this sense silk is more like wool, which also must be spun and assembled, than like hides or meat from large animals, which must be broken down or disassembled to be useful. However, the individual silkworm's cocoon is a one-time resource, and the unwinding of the thread from the cocoon is also a disassembly operation. But the worm's life cycle is so brief and substitutable, it doesn't matter much whether it offers replenishable or one-time resources. So we see another slide in the definition of domestication: The rapid production of silk blurs some of our original distinctions, like one-time versus replenishable resources and assembly versus disassembly.

Citrus killer-ants and silkworms are quite small for domesticated species. Have we domesticated any smaller organisms? Yes. Bacteria have been used indirectly for a very long time in the production of cheeses and yogurts, and they are now a rapidly growing focus of "animal" husbandry. Bacterial uses include the treatment of human waste in sewage disposal plants and the production of vaccines and other drugs through genetic engineering. The first patent for a microorganism was given in 1981 for a composite bacterium with the capability of devouring petroleum molecules that had spilled. It has not been field tested because of the fear that it might spread, but the idea behind it is certainly epoch-making: designer microorganisms that are tailor made to eat and break down specific toxic substances.

The generalization and extension of domestication to ever smaller organisms with brief life cycles must be considered one of humanity's great invention paths. Notice the slide in the concept of domestication again. We have moved from animals to insects and now to bacteria. In each case the concept of domestication readily applies, if we are willing to let it alter slightly.

Can we generalize even further—can we domesticate organisms smaller than bacteria? If we are willing to let the concept of organism slide a little more, so it includes viruses also, the answer is Yes. The discovery of viruses as a distinct class of "organism" may be attributed to Pasteur's work on rabies. He and Koch had established firmly the germ theory of disease, the notion that many illnesses are induced by bacteria. Pasteur then began work on rabies and found that, unlike bacteria, its causative agent could not be seen under a microscope, nor filtered out of a solution, nor cultured in vitro. Nonetheless, Pasteur continued to apply the logic of germ theory that specifies a miniscule foreign agent as the cause of the disease. He established that rabies was readily transferred via the tissue of one animal to another. He then set out to attenuate or weaken the virus, as he had done when developing vaccines for chicken cholera and anthrax, so a rabies vaccine could be developed.

At the right level of abstraction, the attenuation process for the rabies virus resembles domestication, a taming of a dangerous form so that we may use its desirable characteristics. The breeding and range of the virus are restricted, and the taming process begins. Exposing to air the spinal cord of a rabbit that has died from rabies leads to a weakened rabies virus. The attenuated or tamed virus, when used as a vaccine, now produces antibodies that provide protection against the naturally occurring virulent forms of rabies.

Seen in this light, attenuation is simply varying or changing the strength of a virulence variable. It is the virus equivalent of breeding animals for shorter and safer horns. The development of the rabies vaccine is also an analogical borrowing of the vaccination framework from its successful deployment in the prevention of bacterial diseases. But it was a large and skillfully executed leap, because the rabies agent was obviously quite different from that of bacteria; rabies could not be seen, cultured in a dish, or filtered with the then existing methods. Pasteur's genius was guided by still another and perhaps more fundamental framework, his germ theory of disease. Altogether, a beautiful confluence of frameworks and analogical reasoning.

The Transgenic Mouse and Its Context

I want to end this chapter by placing the transgenic mouse in the context that it deserves, one of myth, evolution, domestication, and science. We need myth because, properly formulated, it is the leading edge of our imagination. A good myth is like a fine mirage: it is in front of us and directs our path and our reach. I can't yet see the shape, but I believe the transgenic mouse itself is the raw stuff for new myths. Its existence makes the imagination vibrate, like that mirage over the desert—and like the Sorcerer must have done for our distant ancestors.

Domestication itself turns out to be a surprisingly general idea. Many creatures, particularly those of the bacterial and viral worlds, await a taming hand. Domestication allows for a much more rapid change in creatures than raw evolution does. The contrast between evolution and domestication is the difference between selection based on natural forces and selection based on purpose. By following the path of defined purposes, in a mere few thousand years, humanity has changed the configuration of dogs from wild wolflike forms to the hairless Chihuahua and poodle. Whatever you may think of these domestic forms, the degree and rate of change they exemplify are amazing compared to the usual drift that accompanies natural selection and evolution. The changes have been possible largely because people have had specific goals in mind, like miniaturization, and they have carefully and efficiently selected with these goals in mind.

However rapid the changes due to domestication, they are slow indeed when compared to what can be done with genetic engineering. Of course, what makes genetic engineering possible is scientific knowledge. Now, for the first time, instead of trying to alter biological forms based on surface appearance, it is possible to manipulate the underlying unit of heredity, the gene. Indeed, this is one way in which science helps invention: determining the underlying units. In addition, it helps by finding the principles that govern the combination of those units. These together, units and principles of combination, make possible sophisticated invention and genetic engineering. And even nucleotide synthesis it is now possible to create genes that have never existed before.

There is another way in which the invention of genetic engineering differs from evolution: The inventor may try to breathe life into a stillborn product in the hope that with enough fine-tuning it will eventually spring to life, survive, and thrive. In contrast, evolution requires that each step of development be successful, that it have immediate adaptive value.

Invention also strikingly differs from evolution because the evolution of complex forms lives in a *hierarchy* and invention in a *network*. In evolution the species that form the different branches in the tree of life do not readily mix. Platypuses and pigs don't play and mate in the game of natural evolution but they may in a genetics lab. They may do so because invention operates in a network in which extremely remote nodes come together and mate. In the broader arena of invention, dissimilar forms like trees and ducks mate in the same invention to produce a wooden duck decoy. The joining or mating potential of different inventions in a network of things is far greater, far more widespread, than that of complex life-forms in an evolutionary tree. In principle, any invention can mate with any other invention (the issue of success is another matter), but with evolution a given species can usually mate with only its own kind. (The microbe level is much more like invention, where genetic material mixes much more indiscriminantly, but that is another long story.)

The same point is brought home with simple combinatorial arithmetic. We have seen that for inventions, the number of ways of joining different things pairwise increases on the order of $N \times N = N^2$, a potentially very large number of pairings. And *not one of these pairings can be ruled out a priori as useless or unworkable.* But for different species of higher organisms, outside the genetics lab, the number of possible species pairs that can mate is close to zero. When those same diverse organisms are brought into the genetics lab, any part of the network may be paired with any other, again yielding the exponential explosion of an order N^2 possible pairings.

So matings in the genetics laboratory are really biological inventions. Not surprisingly, biological change can be accelerated when organisms are deliberately invented—that is, when remote pairings become possible. Today's genetic engineers are treating the genetic code as an alphabet for writing biological themes composed by humans. The key variations are produced by experimental techniques that allow for the modification or addition of specific genes. In contrast, the stray mutations that survive natural selection and selection by traditional animal and plant breeders are the result of weak shotgun operations compared to single-gene manipulations that produce rapid and precise variations. The range of organisms and resulting products that can be "domesticated" by gene manipulations will likely astound us all.

Joining the methods of invention to the life-forms of biology must rank among history's most revolutionary invention paths. Through the use of science and invention, biological evolution may, for the first time, approach cultural evolution's rate of change. The combinatorial arithmetic for the number and diversity of things that can come together is one reason why technological and cultural change is so much more rapid than biological change. Of course, other factors forge the explosive nature of cultural and technological change, including the time span of a biological generation, which tends to be longer for humans than for many of our inventions. Nonetheless, in the world of invented forms, virtually anything can be combined with anything else. For good or ill, the possibilities are innumerable.

So what will that invented world of the future be like? How can we go about educating ourselves and our children to the ever-expanding possibilities of an invented world, when those possibilities are fraught not only with opportunity but with danger? Because we know neither the destination nor the direction of our journey, we are in for surprises.

The mythic image of our age may well be that of a white domestic mouse, with small red eyes and a huge breast tumor from a human cancer gene, standing on the shoulders of the Sorcerer. If only we could determine the direction of its gaze, then we would know whether to be frightened or expectant.

Epilogue

Invention Through the Looking Glass

Understanding invention is best realized by looking back and looking ahead. For the several years that I have been groping with the nature of invention, I have had an image that helps me. Like Alice in Lewis Carroll's *Through the Looking Glass*, I gaze into the mirror of invented forms and see myself as a speck on the crest of a moving wave. As that speck, I frequently change views, alternating between looking back—at history and earlier understandings—and looking forward—at the implications of invention and the principles that underlie it. Because you have been on the wave with me, together we need to take one more look back along the path traveled and then one more brief glance ahead, both looks while perched precariously on the crest of invention's ever-changing wave.

Hindsight

We have seen that Invention begins with need, branches into want, and flowers with possibility: We need knives to gather and process our food, we want to ascend to the clouds in hot-air balloons, and in stirring our test tubes, we see the possibility of new life-forms.

We have examined invention in children and in experts. From one view, a great gulf separates the efforts of the child inventor and Leonardo. Still, beneath the surface, many common currents work their way. These show up in our descriptive framework, where we see diverse inventions sharing categories like purpose, principle, parts, and evaluation. That framework allows us to see connections between cooking pots and hot-air balloons.

We have abstracted heuristic principles that underlie the shaduf and the nuclear reactor, the cartridge and the TV dinner. These heuristics, when applied to existing inventions or their descriptions, help us generate new ideas and track them through the vast spaces of possible invention. Many of these heuristics focus on single inventions, but some of them help us combine or join together very different inventions in a pattern of broader integration.

We have seen the uses of natural materials, like the buffalo and its parts, that serve to define the civilization of the Plains Indian; then the synthesis of diamond from peanut butter; the development of new materials like Kevlar with a strength that exceeds any fiber found in nature; and finally the beginning of a quest for smart materials.

We have explored interfaces that connect us to our artifacts: the slow evolution from the handle on a stone knife to the multihands handle of the two-person saw, from the violin bow to the keyboard of the computer-controlled synthesizer. We have studied the art of containment, ranging from the cooking pot to the electronic prison. We have seen procedural inventions that span methods for acquiring goods in a supermarket to the ways of assembling them, from handcraft to assembly line, to the ways of maintaining them by self-test or by remote test.

And we have tracked through time the evolution of transgenic life-forms: first those like the Stone Age Sorcerer that early on spring from the rich operations of abstraction and joining that the human mind enjoys, then to the practical process of simulating life through decoys and scarecrows, still later to the fine-tuning of real life through centuries of domestication and selective breeding, and finally to genetic engineering based on the manipulation of individual genes. Our ending point is a mythic mouse with a human breast cancer gene, standing on the shoulders of the Sorcerer.

Clearly, invention is more than the front-end of technology; invention is also a way of thinking. To study invention is to discover hidden intelligence all around us. And underneath that intelligence we reveal powerful mind tools, invention heuristics and principles often used intuitively and unsystematically. These mind tools are easiest to see in the embodiment of a concrete invention, but they are not restricted to technology or any one subject matter. They travel on their own waves through many subject matters and across great sweeps of time.

At bottom, this is what invention is all about: moving back and forth between artifact and principle, back and forth between hand tool and mind tool.

Foresight

Not long ago, I met a physicist who told me that he had never had a manuscript rejected by a journal editor. He was proud of this, but I found his statement troubling and was not sure why. Perhaps it was envy, because I have had my share of rejections? The more I thought about it, the clearer the answer became: He had not been reaching far enough. If we do not fail from time to time, we have no real idea how far our abilities stretch, and we are not risking enough for them. We as creators are not living up to what is possible.

Why is it important to play on the edge of imagination to be creative? After all, the inevitable consequence of regularly playing on the edge is failure. Perhaps the genius can play easily within the borders of his or her ability and still produce remarkable work? Yet even the person with undeniable genius is most likely to make a great contribution by stretching a bit. And for most of us, to do something really inventive or creative, we have to stretch a long way.

As I thought more about this, I began to see the creative behaviors of many of my students in a new light. Often they are not living up to their ability, not because of lethargy but because they are afraid to stretch beyond themselves. That failure to stretch is due to a fear of failure, embarrassment, bad grades, and a host of related consequences. The fear is not without basis in the everyday world. People and events are constantly evaluating our expressed thoughts and actions. A miscue can have disastrous effects.

How can these two ideas be put together—the need to stretch our minds to the fullest and the need to avoid the adverse consequences of failure? What I now try to do with my students is to have explicit times when we can present our ideas with safety and other times when we know that the real world of evaluation has crept in. These safe times are not brain-storming sessions, because even a wild idea may be worked out in some detail and have many evaluative criteria taken into account during its formation. Nonetheless, during these times the inventive imagination is encouraged to work its way.

This strategy has relevance to corporate and school settings. Often we do not discuss our wildest ideas with colleagues. For good reason. Usually we are evaluated by people in our own department, and we don't want to sound off the wall. So we often need to find other settings where we can try out our wild ideas without fear that the colleagues who evaluate our raises and promotions will be watching over our shoulders. A good setting for stretching the mind is a friendly and informal cross-disciplinary group, where no costly evaluation is looming; the members of the group are chosen from administrative units other than our own. My favorite setting of this type is a lunch with one or two colleagues from *other departments*. In this setting we can often work at the edge of our ability, knowing full well that we will have many failures. In these sessions, we try to be quite critical but in a humane and constructive way.

Minimizing the threat from evaluation is not the only advantage of such a voluntary group. The interdisciplinary idea is fully as important. In addition to playing on the edge of our ability, we need to play on the edge of our discipline, where the boundaries are. Everyone argues for interdisciplinary cooperation, but the successes are fewer by far than the failures. Some explicit mechanism or procedure is needed to make boundary crossing work.

In this light, one of the more interesting procedural inventions I know is the practice of *sconcing* at Cambridge University in England. I was told about it by John Wild, one of the first scientific investigators and inventors of the clinical use of ultrasound in which the body is scanned with sound waves that echo differently off normal and pathological tissue. For this work, Wild recently received the Japan Prize, the Japanese equivalent of the Nobel Prize. The roots of his work go back to student years at Cambridge University in the 1930s and the practice of sconcing.

The practice worked this way. Students lived in separate colleges, and they were required on five nights of the week to either invite a guest to their college for dinner or to be a guest in one of the other colleges. During these dinners, it was forbidden to talk to someone in your own discipline about your discipline. That was shop talk, physicists talking to physicists. However, to talk to a person in another area about your discipline was approved, perhaps physicists talking to biologists.

If you violated the rule by talking to someone in your own discipline, then you were sconced. Sconcing consisted of being fined, with the fine paying for a huge chalice of ale brought to the table for everyone to drink. Because no one wanted to be fined, Wild says, the interdisciplinary talk was very intense. He credits it to developing the background and breadth he needed for working in clinical ultrasound—the bringing together of physics, engineering, physiology, and medicine. The sconcing procedure may not be exportable to settings where the ale does not flow freely, settings like a public school. However, the more general idea underlying it certainly is exportable: Find better ways to bring people of different disciplines together to share ideas.

While sconcing facilitates interdisciplinary thinking and the crossing of boundaries at an informal level, it may not go the extra mile in facilitating active coordination. To do that in an existing bureaucracy, we need ways of tearing down the Berlin Walls put up by administrative units, or at least a way of making those walls more permeable. Recall the Sony Walkman example in which Sony's retired chairman walked between department boundaries in his personal effort to put together the various ideas developing in different labs. This is a general and potentially quite powerful idea that is readily exportable to many organizations. All it needs is the institutionalization of using older, respected members of an organization to do the walking.

Whatever the considerable potential of the Walkman model for linking labs and ideas, it is still local, still within an organization. Ultimately, we seek a still broader group to play on the edge with. We need coordinating groups, spontaneously formed by mutual interest, that are global in the literal sense of spanning the world. The groups should be capable of forming and dissolving on the fly or persisting indefinitely; and they should be low in cost and accessible to everyone. In this light, computer networks and bulletin boards look promising. Currently they work well for problems that do not require intellectual ownership, problems like coping with a particular brand of computer.

When intellectual ownership of ideas is an issue, then restrictions on the flow of information become a factor. There is a way around this within a company or unit, by using the Walkman model extended to the computer network. I'm not sure yet how to do it on a broader basis where we need to worry how to credit individual contributors. At the rate computer networks are developing, however, I will be surprised if methods to handle this vexing problem are not already forthcoming. Perhaps posting an idea and one's name on a computer bulletin board is a good way of nailing down priority?

The computer network idea is at once highly general and powerful because the agents we communicate with (fellow humans) have a much better model of human understanding than current computer programs and artificial intelligence. A good human correspondent knows what we need to know even if we don't know ourselves. Such a correspondent can help us play on the edge of our abilities and at the same time provide a net to keep us from a disastrous fall.

The ideas of sconcing and a wandering Walkperson seem especially promising, whether worked face to face or on a computer network. Let's see if we can work up to a more general application, such as a classroom of children.

In a classroom, I believe that historically early and momentous inventions are the best path to learning about technology, because they are easier to understand than their complex modern embodiments, and they better reveal the formation of fundamental ideas and decisions. And invention is the best departure point for the understanding of technology because it is concerned with ideas and principles that quickly cut through the glitz and dazzle of the latest gee-whiz technology that we see in the popular media. Nowhere are important ideas and principles better embodied than in the early inventions of a technology. Once principles are understood, then more modern ideas can be introduced. For those classrooms or countries that cannot afford computers and science kits, a rich set of culturally meaningful materials is always available all around one for the study of invention and technology. In its own way, the potter's wheel tells us as much about invention as a nuclear reactor.

In any study of invention and technology, numerous opportunities will present themselves for also studying the nature of measurement, decision making, search, and experiment. More generally, the development of thinking skills should be stressed. Many opportunities should be available for the development of writing and speaking skills. This view is very much in keeping with the American Association for the Advancement of Science's approach to instruction about technology, as described in the *Project 2061* curriculum. This curriculum is an important thrust because traditional instructional emphasis on science and math does not cover much of what goes on in technology, either in thinking style or in social impact.

What should be done in a good curriculum to teach our youth about invention and technology? Here are some ideas, no one of which is likely to be expensive or demanding of unusual resources.

First, we must get beyond the glitz of technology into principles and mental representations. While I have not done so here, because it is another story, concepts from mathematics and science should be regularly introduced in the invention curriculum. In this light, I recently read the *Project 2061* proposed curriculum for mathematics. To my surprise, the concepts to be emphasized are not computations and proofs but what the professional mathematician does: abstraction, representation, inference drawing, application, and generalization. In invention terms, we have talked about abstraction and parsing, a descriptive framework as a way of representing inventions, heuristics as a way of generating ideas, applications, and certainly generalization. The convergence here between the processes of mathematics and invention is gratifying and needs further elaboration. I believe that this modern style of thinking about mathematics instruction will make for easy bridging with the framework for invention that we have built.

Second, we must see invention and technology in a broader context, one that not only develops inventive thinking but that also shows social connections, another important part of invention that we have intentionally neglected. Come with me now to see what a typical open classroom for the study of invention will be like. We enter the class and you are surprised by an absence of fancy electronic gear. Young children are looking at pencils and erasers to note their components and materials. They are playing What-if games by trying to imagine how the pencil will work if the parts are rearranged or other materials substituted. Then they are trying to guess how the pencil is manufactured, the different ways of making a pencil. What kind of manufacturing equipment is necessary. How difficult would it be to make an ordinary yellow wooden pencil from scratch? Along the way, they are learning about describing the pencil, using categories like Purpose, Principle, and so on—all to establish a descriptive framework for invention.

An older group is studying bread-making, a project that will connect them with an ancient history and the latest manufacturing automation. They have baked a single loaf, carefully measuring the ingredients and the time for each step—all to study discrete process manufacture. Now they are looking at the ingredients and time for making several loaves concurrently—to study batch-processing manufacture. Later, they will take a field trip to a modern bakery where they will see automated manufacture that never really stops—to study continuous mode manufacture. Back in the classroom they will look at the development from discrete to batch to continuous manufacturing, paying close attention to how the manufacturing processes at home gives way to those of an automated factory and simultaneously parallel operations.

Then they will be ready to simulate an assembly line in the classroom. Paper constructions with gluing and stapling handled in discrete or batch mode will work well here. Then a field trip to an assembly line is planned. Some discussion of the assembly line's social effects will be in order. Perhaps a worker will visit and talk about life on the line. Economics of different assembly procedures can be introduced and the relative advantages of different manufacturing modes can be listed and discussed.

At the junior high or high school level, students are recreating the Wright brother's wind tunnel experiments. Concepts like lift, gravity, drag, and thrust are introduced and measured. Similar units on the development of the telephone and the electric light will introduce fundamental technologies. A feeling for the importance of these inventions will be communicated by living with phones unplugged and the electricity turned off for an evening. Essays will be written describing how one's life changes when there is no phone or electrical service. Contemporary complexity will also be studied. If the school system has computers and networks, students can trace these inventions back to earlier ideas.

Invention and technology, however, must be more than understanding great ideas of the past. One of the essential conclusions from studying invention is that no one can predict far ahead which ideas will come together. We have seen this over and over, from the combination of the inclined plane and geometric packing that produce the screw, to the fundamentally different ingredients that come together to yield the phonograph, to the combining of animal parts that resulted in the early Sorcerer and the much later mouse with a human breast cancer gene. To catch this exciting element of invention, students must have the opportunity to play invention games on the edge of their ability and imagination—and to play some wild cards. How can we implement wildness in our curriculum, as advocated by Perkins?

The earlier ideas of sconcing and the Walkman aid us. The essence of sconcing is having people talk to one another about their different projects—all with the idea of exploring interesting and surprising connections. While an elementary school cannot pass a tankard of ale around the cafeteria table, something similar can take place.

Imagine a game in which some special tables are set aside in the cafeteria, and once again children get to occupy these tables by inviting someone from another invention project. The conversation follows the same rules as at Cambridge University: You cannot talk about your own project to someone in your own group; you have to talk to people working on other projects, sharing ideas between projects. Violation of the rule may mean sacrificing your dessert and everyone else getting an extra dessert, in a manner similar to Cambridge's use of ale. (Obviously, someone more expert than I about children's games needs to set this up!) The idea is to generate ideas that blend the inventions of different ongoing projects. Perhaps a contest can be set for writing the most interesting proposals combining ideas across projects?

In a related manner, the concept of Walkman or Walkperson is implemented. The Walkperson will be a particularly skilled older student who walks between the different projects with the idea of combining ideas. How can the activities of one group help those of another group? If this is too difficult a task for students, a retired engineer or scientist can be enlisted to help out.

The objective of both the sconcing and Walkperson procedures is the development of a wild idea. The participants will use their free reading time to work on the edge of their idea and their ability. Essentially, this is a classroom implementation of the corporate lab's Skunkworks in which an elicit idea is worked on without corporate sanction or with the immediate managers looking the other way. In the corporate Skunkworks people believe enough in an idea to give their all for it, even if it is a bit off the path of company objectives. It is a way of keeping industrial labs innovative, and perhaps it can do the same for the classroom study of invention.

Just as the good manager will look the other way as long as the Skunkworks project is remotely related to corporate objects, so will the good teacher play the game by looking the other way. The only requirement of students is to periodically report on their progress so the moving wave of the idea can be charted. (Safety requirements and the like can be handled by the retired scientist or engineer.)

The Skunkworks takes us back to our earlier idea of play-driven invention: combining interesting ideas freely—but seldom randomly—in a supportive or neutral environment, all the while guided by invention principles. It is the culmination of the independence and self-initiative that we want to foster in children who live in a technological world.

Alice, in Lewis Carroll's *Through the Looking Glass,* leads us to a vanishing point. When Alice looks in the glass, she sees not herself but an unknown world of fantasy where everything is reversed or backward or transformed in scale. Of course, she really does see herself in the looking glass. It is her own imagination that is mirrored in the glass.

Likewise, when we look closely at the artifacts and processes about us—many different looking glasses—we can see more than a physical presence. We can see deep invention principles in simple acts. These invention principles reveal a hidden human intelligence, a reflection of our own mental processes. When we understand the principles and hidden intelligence of invention, we see—mirrored in the peeling of an apple—the way our own minds work, the way in which not just airplanes lift and turn but how imagination soars. By understanding principles behind simple invented forms we glimpse the halting steps, graceful turns, and exhilarating leaps of the inventing mind. And with the anticipation of Alice, we enter an invented place, a world of our own making.

Notes

Preface

For anyone who wants to examine the social aspects of technology—the part I am intentionally neglecting—I recommend: Bijker, Hughes, and Pinch (1987); Mokyr (1990); and Westrum (1991). Some people will claim that the process of invention cannot be separated from the social. In an ultimate sense that is true, and capable people are now examining the social fabric. But many invention principles can be understood without recourse to the larger social fabric, understood perhaps only by looking away from that fabric toward the details; very few people are studying invention from this view. It is the view that I pursue here.

A different approach to creativity, one based on the biography of individual creators, is that of Wallace and Gruber (1989). And still another is the multiple intelligence approach of Gardner (1983). All these views of creativity are valuable, but they do not tell us much about specific inventions, the path we will follow here.

Prologue: Peeling an Apple

For stone tools: M. Leakey (1971). For the domestication of the apple: Zeuner (1954a).

1. A Context for Invention

For the requirements of patentability, see U.S. Department of Commerce (1985) and Schneider (1988). The mouse with the human breast cancer gene is described in the patent of Leder and Stewart (1988).

The idea of an aggregate intelligence or "folk mind" is an ancient one, often discredited and frequently reconstituted as needed; I have taken it from Wundt (1900–1920).

Figure 1.1, "The Coffeepot for Masochists," is from Carelman (1984, p. 43). Carelman's book is a magical mystery tour through a world invented by a fine prankster. It is in French, but anyone can understand the pictures. I first became aware of it as the cover illustration of Norman's book, *The Psychology of Everyday Things*. In fact, Norman (1988) is a very useful complement to the approach taken here. His view is the discipline of human factors, the fit of human and machine to one another. When the human factor is neglected, a common design procedure is to develop the machine first and then, as an after-thought, try to get it together with humans. Anyone who has a VCR with a blinking 12:00 will understand this—and the importance of considering the user from the beginning.

For an extensive treatment of design, see Simon (1981) and Suh (1990).

For additional distinctions between invention and science, see Weber and Perkins (1992). On the relation of invention to science, technology, and the visible, see Tweney (1992). It is with some hesitation that I have tried to draw distinctions between invention and science. Many people will say there is no difference, that the two are opposite ends of some continuum. If so, I ask what is the nature of that underlying continuum? Claims to unity must be backed up by better arguments than those usually given. One common argument is that science discovers truth while invention and technology apply that truth. This is a claim that will not stand scrutiny because invention/technology is so much older and more universal than science—people have developed many important technologies without waiting for science. I think the distinction between science and invention is more complex than anyone has yet characterized. Certainly, my attempts here are subject to their own problems.

On problem solving, see Newell and Simon (1972), Simon (1981); and on the recreation of important scientific discoveries, Langley, Simon, Bradshaw, and Zytkow (1987).

A good discussion of art as a creative process is given by Winner (1984).

For reviews of the creativity literature: Kim (1990), Perkins (1981), and Weisberg (1986). A couple of classics are Osborn (1963) on brainstorming and Zwicky (1969) on morphological analysis. A critical review of brainstorming's claims is provided by Weisberg (1986). Essentially, brainstorming groups do not reliably generate more or better ideas than a comparable number of individuals working alone. Nonetheless, brainstorming can be very useful for promoting social cohesion and getting people to work with one another. The Zwicky idea of morphological analysis is to generate all possible combinations of a problem and then pick out the best answer. That works fine for small problems, but the development of any moderately complex invention is likely to generate so many possible combinations that the method is inapplicable in practice.

2. Novice Invention and a Problem-Based Diary

I wish to thank the following individuals for their help with materials for the section on children's invention: Irwin Siegelman and Barbara Taylor of the

Weekly Reader; and Audry Benson, of the U.S. Patent Model Foundation. For the section on children's inventions, I drew on the following materials: *Invent America Newsletter;* United States Patent Model Foundation, February 1987; *Weekly Reader,* 1986 and 1987 award winners, news releases, undated.

I also wish to acknowledge the efforts of three dedicated teachers who put together the Stillwater Invention Program and taught me much of what I know about children's invention: Virgalee Reed, Beverly Riggs, and Susan LaForge.

Useful books for the inventing child: Macauley (1988) and Shlesinger (1987).

3. Expert Invention and the Turn of the Screw

The idea of invention as play-driven is strongly argued by Basalla (1988) who writes convincingly that very little of invention is based on necessity: Most needs can be satisfied any number of ways, and almost always a range of inventions and technologies can be selected from when appropriate; moreover, many inventions are not directed toward any obvious biological need.

The screw as a fastener did not come into wide use until the mid-nineteenth century (Usher, 1954). The difficulty of making screws would have prohibited their early use as fasteners, even if the idea had been present.

On Archimedes and the water screw: Drachmann (1967), Usher (1954). Figure 3.1 is based on Drachmann (1967). Some writers (Gille, 1986) claim that the screw, in some form, was invented earlier by Archytas. Fortunately, our concern is not with definitive historical truth but with constructing plausible mental paths to invention, so we need not know the actual inventor of the screw. Indeed, it is likely that multiple inventors were involved over a long period. To give either Archytas or Archimedes credit for it is probably an oversimplification, even if it makes writing about it easier. Forbes (1956) claims that the first water screw was not hand operated, but, instead used a treadmill. The treadmill proposal is doubtful, however, because the screw enters the water at an angle that would make it difficult to fit a treadmill to. The fact remains that at some time the water screw acquired a handle, and we are licensed equally to trace the developmental path of an invention whether that path is entirely within one mind or crosses many minds and many centuries. Indeed, almost all important inventions are the work of multiple minds. But once we extract principles behind their development, it is possible to incorporate those principles into the individual mind, thereby giving us as individuals a leg up on the inventive process.

Dean Simonton suggested to me that the staircase is not really an inclined plane but a series of rises and runs. I found this an interesting notion, but if you look at people on stairs it is clear that a good deal of diagonal movement is present; this makes stairs an inclined plane. Think of it this way: a person has a center of mass and that center is moving diagonally. I communicated with Dean about it again, and we are both agreed that the fit of physics and real world is not very precise here. He made the additional interesting suggestion that whether stairs are a ramp or a step function depends on the kind of user, a bounding teenager or an infant making a series of discrete up-and-across movements. Of course, the true advantage of stairs over a single leap to the top of the building is based on the relative power requirements, that is, work per unit time. The required power is certainly much greater for the superperson leap.

On the idea of mental play as a basis for invention, see Blackwell and Helkin (1991) who discuss play in the context of mathematical invention. The idea of Archimedes' water screw has been generalized to the movement of grain with an auger. On a large scale, this was done by Oliver Evans who produced a close approximation to an automated flour mill in the 1790s. The grain screw is commonly used in grain elevators.

Drachmann (1963) talks of the early history of the screw. The screw press goes back before Heron, who documented its use, to probably the first century B.C. (Usher, 1954, pp. 126–128). For the construction of the pyramids using an inclined plane, see Lloyd (1954).

On Leonardo: See Pedretti (1987) for the sketch of the helicopter. The sketch here is very similar, with some simplification, and my thanks to Maria Ast for rendering it. The story of the child's toy also is in Pedretti. Leonardo's quote on the helicopter is from MacCurdy (1958, p. 500; italics and comments added).

The olive press is based on Usher (1954, pp. 126–128). Leonardo also sketched a printing press based on a screw through a wood frame (medium). The suggestion that the helicopter may have come through the olive press by analogy is my own speculation; I know of no evidence other than the striking correspondence between the two—and the circumstantial evidence that Leonardo was fascinated with screws of many kinds.

The section on the Wright brothers is based on the work of Crouch (1981, 1989, 1992) and also the Wrights' own account (MacFarland, 1953). The question of "Why Wilbur and Orville?" was raised by Crouch (1989b). It intrigued me greatly, and I have attempted to answer it by drawing on Crouch's work as well as some of my own ideas. The Wright's patent was not for flight as such, but for the mechanism of controlling flight by warping the wings to make them function in what I interpret as a screwlike fashion.

The idea of following the screw through the inventions of Archimedes, Leonardo, and the Wrights is my own.

4. Describing an Invention

In my technical work, I use a descriptive system that more closely approximates what the people in artificial intelligence call a *frame description* (Lenat, 1978, 1983; Minsky, 1975). For the adaptation to invention, I am indebted to many long luncheon conversations with D. N. Perkins whom I met while I was on a sabbatical leave at Harvard in 1984–85. I am well aware that AI people would find the descriptions loose and not specific enough for programming. That does not bother me because I am concerned with human rather than machine communication. See this argument in more detail under the later section dealing with power-generality tradeoffs. References to using framelike data structures for describing invention include: Perkins (1987) who uses the categories *purpose, structure, model,* and *argument* in the context of instructional design; Perkins and Laserna (1986); for explicit application to inventions, see Weber and Perkins (1989), Weber & Dixon (1989), and Weber, Moder, and Solie (1990). Other investigators have also picked up on the idea of a frame description for invention. For example, Carlson and Gorman (1992) use the idea in conjunction with the idea of mental models to explain the invention of the telephone by Bell and Edison. They also supplement it usefully with the idea of mental models. I believe mental models are a powerful addition, but they usually depend on a live inventor or a dead one who left extensive notes—a rare occurrence.

For *scripts*, see Shank and Abelson (1977).

For a brief history of the eating fork, see Panati (1987). It is surprising that the fork is so recent; the knife and the spoon are much older.

5. Evaluating and Comparing Inventions

For a critique of brainstorming, see Weisberg (1986). For an architectural approach to evaluation that is close to my own, see Alexander (1964). Some of the comparative ideas are from Perkins and Laserna (1986).

A more detailed exposition of evaluation and comparison is presented in Weber, Moder, and Solie (1990), where the notion of a constraint table is introduced and made explicit.

The evaluative criteria in number theory are from Lenat (1978). Those for the longbow and firearms are from Esper (1965).

6. Understanding the Created World

The Fulton quote is from Phillip (1985, frontispiece).

The soup ladle and doorknob examples are in the blurry zone between invention and design. I suspect that the difference between a general-purpose serving spoon and a soup ladle is large enough to consider them two different inventions, but not everyone will agree. Some people may wish to say that both are simply design variants of a spoon.

If the doorknob example interested you, look also at Norman (1988) who examines a variety of human interfaces from a different perspective: what is wrong with them.

7. Heuristics as the Engine of Variation

For sophisticated Darwinian models of invention: Basalla (1988), Campbell (1960), Simonton (1984, 1988). Chance is certainly important in the generation of inventive ideas, but to anchor it more firmly as an explanation requires a better specification of the units of thought that are undergoing chance variation. Thus a good example of chance as explanation comes from current speculation on the origin of life (Margulis & Sagan, 1986). The chemistry of the primordial soup is known (the units or constituents), and when bombarded with simulated lightning (the operations), the chemicals spontaneously form organic molecules (the result). After the fact at least, all of this is pretty much in the context of well worked-out chemical theory. When it comes to explaining inventive ideas by chance, we have no comparably worked-out set of units, operations, results, or theories. This does not say that we will never have them, just that the rush to explanation based on chance may not buy us as much as we wish.

For the idea of heuristics applied to the field of mathematics: Polya (1947–54, 1954). On heuristics and problem solving, Newell and Simon (1972). On heuristics and scientific discovery: Langley, Simon, Bradshaw, and Zytkow (1987).

For the haystack metaphor elaborated in the context of invention: Perkins and Weber (1992). The idea of looking for inventive ideas in a search space is closely related to the problem-space conception of Newell and Simon (1972).

The power-generality tradeoff: Weber and Perkins (1989). For human intelligence versus machine intelligence: Perkins and Martin (1986); Weber and Perkins (1989).

My approach to heuristics differs somewhat from that of Simon and his collaborators. First, I do not use computer simulation, so what I am doing is much less rigorous than Simon's approach. By the same token, I often study ancient artifacts for which there is no corresponding written or verbal record, the very requirements of Simon's protocol analysis. In general, the much greater flexibility of human than machine intelligence provides the basis for my approach, as does the fact that my primary purpose here is to communicate with fellow humans rather than computers.

Second, I do not claim that invention is merely problem solving, as Simon does for scientific discovery and might well do for invention. As has been said, "god is in the details," and the details of invention seem to continually suggest new principles that go beyond any present statement of what is merely problem solving.

Third, the topic of invention seems to present a number of principles that distinguish it from conventional problem solving or discovery in science, and invention often presents interesting and enduring examples that are much simpler than the topics studied by Simon and his collaborators. For instance, it is easy to find inventions hundreds of thousands of years old—knives, axes, and the like—that are still important. I find it difficult to see what computer simulation or the claim that these are mere examples of problem solving will add to the account of these artifacts. There is no substitute to rolling up the mental sleeves and trying to understand these artifacts at their own level. Having stated these caveats, it is obvious that my approach to invention owes a large intellectual debt to the work of Simon and his associates.

The Einstein quote is from Hadamard (1954, pp. 142–143); the Planck quote is from Planck (1949, p. 109); and the Rabinow quote is from Rabinow (1990, p. 240).

8. Single-Invention Heuristics

The idea of the variable is very important, and it is further illustrated by Whitehead (1948) who traces the development of the concept in mathematics. I think there is a close relation to the use in math and the potential use in invention, but not everyone will agree.

The idea of a *slide* is closely related to Hofstedter's (1985) concept of *slippage*. However, sliding may be more deliberate than slipping.

Simon (1981) uses a concept called *satisficing* in a way that is related to what we have called fine-tuning. I just thought that fine-tuning is a more descriptive term and more widely used in the invention context.

The shaduf: (Drower, 1954). The conceptualization of variance control is my own.

The factual material on the Walkman came primarily from Nayak and Ketteringham (1986).

"People like myself . . . ," "So if you . . .": (Nayak and Ketteringham, 1986, p. 58).

The idea of the *Walkman* having a double meaning—including that of a person integrating the work of different labs—is my own; so too is the use of the Walkman to illustrate feature addition and deletion as a common form of invention.

Another common example of invention by feature addition may be the chair as indicated by the sequence: one-legged stool, three-legged stool, four-legged chair, an added back, and then added arms. Many other examples abound: jackets with zip-in or-out linings, or zip-on or zip-off sleeves.

9. Multiple-Invention Heuristics: Linking

The factual material on the Post-it Note is from Nayak and Ketteringham (1986), as are the quotes from Silver (p. 58). The New Purpose heuristic is my own abstraction.

10. Multiple-Invention Heuristics: Joining

On bisociation and the role of the unconscious in invention, see Koestler (1964).

Types of joining is based on long hours of staring at tools (Weber, 1989).

On inverses: Weber (1992); Weber and Perkins (1989). For the idea of the eraser applied to the pencil: Petroski (1989).

For parallelism and emergent functions: Katie Harding in the news release for the *Weekly Reader* contest (1986).

Levels of joining: Weber (1992). One of the best ways to study joins is to look at a catalog. Joins can readily be seen in tools, kitchen items, furniture, and a host of other places.

11. Transformational Heuristics

For an early implicit recognition of the idea of tool transformation, see Pitt-Rivers (1906).

The idea of transforming an element to produce physical inventions is my own; it is treated explicitly in Weber (1992); Weber, Dixon, and Llorente (in press); Weber and Perkins (1989). Of course, I make no claim to its historical use; it is an after-the-fact organizer.

Still other tools may be thought of as building on the idea of the transformed tooth. A medieval weapon, the flail, has a set of sharp teeth on the surface of a metal sphere, all attached to a chain and handle. This terrible weapon allows one to whip around an opponent's shield. The flail combines teeth and sphere, with the teeth on the *outer surface* of the sphere. A puzzle for you: Of what possible use or application is a hollow sphere with the teeth on the *inside*, pointing in toward the center? If you wish, you may assume that the sphere will come apart as two hemispheres. Let me know if you think of a good application.

Transforming an egg is based on hours of staring at eggs. I am indebted to my friend Mike Lyons for an evening spent in the rigors of testing the egg's strength by using a bathroom scale—we could then determine the force required before the towel-wrapped egg finally broke. Because that force is so great that no one will believe me, you will have to repeat the experiment with your own egg, towel, and scale.

12. Discovering Heuristics

The material on the wheel is based in part on research done with my student Antolin Llorente on the perceived origins of the wheeled cart (Weber & Llorente, in press). For the historical origins of the vehicle wheel: Piggott (1983); Littaur and Crouwel (1979); Derry and Williams (1961); Fabre (1963); and Neuburger (1969).

Sometimes people say that the wheeled vehicle was so long in coming because in ancient times there were no roads or suitable terrain for it to move on (Basalla, 1988). This explanation seems to have in mind two-wheeled if not four-wheeled vehicles, and it is unconvincing. During the Vietnam War, a principle mode of supply conveyance by the North Vietnamese through the jungle and down the Ho Chi Minh Trail consisted of pushing a heavily loaded two-wheeled bicycle. The Ho Chi Minh Trail was not paved. Even simpler than a two-wheeled vehicle is the possible application of the wheel to one of the oldest conveyance forms, a travois. The travois is a triangular-shaped platform with a minimal surface area skid where the triangle comes to a point. Conceptually, it is a simple matter to attach a single wheel to such a device, indeed much simpler than to build a cart with multiple wheels. It is in the nature of a single wheel that it can go almost anywhere that a person can walk, but I know of no early historical evidence suggesting that a single-wheeled travois was ever used. Unlike the travois, the wheelbarrow is pushed rather than pulled; in addition, it has an attached container. It also seems to be a relative late comer; evidently, its origin is about the third century A.D., due to the Chinese (Needham, 1965; Temple, 1986). Surprisingly, one-wheeled vehicles apparently evolved later than two- and four-wheeled carts.

The material on the needle is based on Weber and Dixon (1989). Other sources include: Bordes (1968); Dennell (1986); Stordeur-Yedid (1979); and White (1986). Dennell asserts that sewing is one of the great unheralded inventions of human history, and I find myself very much in agreement.

The middle-eyed needle is from Stordeur-Yedid (1979). She believes that it may not have been used in continuous sewing at all. Instead, it would have been used in discrete sewing by poking it through halfway, threading a discrete stitch, pulling the needle and thread back out, and so on. If so, it is a false path to the base-eyed needle. The joined fork and awl needle is from Bordes (1968).

13. Applying Heuristics: Inventions After Their Time?

On the hot-air balloon: Needham (1965), who presents evidence for the early Chinese use of hot air to loft empty egg shells skyward. The first human flights in a hot-air balloon were conducted by the Montgolfier brothers: Scott (1984).

I want to thank my friend Mike Folk for a great luncheon conversation about the phonograph!

On the phonograph, see Edison (1878) for what must be one of the great documents in the history of technology. This patent contains the idea of removable secondary information storage and mass copying and production. See also Conot (1979), Jehl (1990), and Wile (1982).

The absence of a mental model or concept as a contributor to the delay of the phonograph is my own idea, but see Carlson and Gorman (1992) who provide a good discussion of mental models and their applicability to invention.

I have not been able to find a drawing of the Scott phonautograph of 1857. A rough sketch of Koenig's version of 1859 is shown in Figure 13.2; it is from Miller (1934). A very helpful source on the early history of the phonograph is Wile (1982) who seems to have done the best job of tracking what Edison really did and when—in contrast to fable, promotion, and legend.

The origins of Velcro are elusive, but a company brochure fills in part of the story (Velcro USA, undated). The key Velcro patents are de Mestral (1955, 1961).

My first exposure to the virtues of Velcro came from working in a theater company behind the scenes. For a rapid change of costume in the dark, there is nothing like it. One fastener I have not been able to find out much about is the snap. It is interesting to speculate on how snaps came about. One dark and stormy night, Dr. Frankenstein was working on hip sockets and. . . .

Some qualifications and caveats on the asserted importance of heuristics require mention. So far, I have argued that inventions after their time are the result of not having, or not systematically applying, appropriate heuristics. But it is possible to have an invention failure for the opposite reason: Using a heuristic or model that is inappropriate. Examples readily come to mind and include bird models of human flight with flapping wings and ox collar models for the horse. Both of these are very bad ideas, which suggests the need for another essay at another time entitled "Great Analogical Failures." So heuristics, especially those concerned with analogies, offer not only great opportunity for guiding the mind through the space of possible inventions but also some hazards if we are too attached to the wrong ones. We need heuristics for telling us when we are applying bad heuristics. That will have to await another book.

14. A Material World

The uses of the buffalo by the Plains Indians: a chart from "The American Bison Association" (undated).

"[A] record-player pick-up cartridge . . .": French (1988 p. 53).

Bamboo and its uses is from Temple (1986) and Needham (1971). Edison's use of carbonized bamboo as an electric light filament is from Friedel and Israel (1987).

The material on diamond is from Amato (1990), Graf (1987), Smalley (1991), and Wentorf (1992). Some informal conversations with Wentorf were also important: The Inventing Mind Conference, Tulsa, OK, November 3–5, 1989.

The material on Kevlar is from Hounshell and Smith (1988), Morgan (1992), and informal conversations with Morgan: The Inventing Mind Conference, Tulsa, OK, November 3–5, 1989.

Polychromic glass: Trotter (1991).

15. *The Interface's Form*

For early stone tools: M. Leakey (1971). For early handles on stone tools: Bordes (1968).

On fuzzy boundaries between categories: Lakoff (1987).

The importance of the stirrup: L. White (1962).

The illustration of the horse collar in Figure 14.5 is adapted from Needham (1965, p. 305). The information on load that can be pulled with the different collars and the dates of use: Needham (1965, p. 327).

The material on musical interfaces is a miscellany of ideas that I put together from hanging around music stores and musicians.

We have touched on a powerful form of device-to-device interface, a computer-controlled synthesizer. More than one musician is concerned about what such technologies will mean for his or her employment. Indeed, ever more of commercial music is produced by synthesizers; and that trend will be accelerated with the linkage to computers. The commercial advantage of the synthesizer over a variety of other instruments is clear—only a single musician must be paid and communicated with. The disadvantage is an economy that will support fewer musicians and provide less musical variety. For more on the social implications of invention, see Westrum (1991).

16. *The Art of Containment*

The information on tea bags is based on staring at tea bags.

The idea of sliding in and out of containment is just a theme and set of variations on the idea of containment. I want to express my appreciation to Margaret and Sydney Ewing for several of the container examples.

"People can reach the food at the bottom of the jar . . . waste" is from *Weekly Reader* (1986, news release).

The body as container: Johnson (1987).

If you have not figured out the mystery container, carefully consider its relation to the fish trap. Let me hear your ideas on this.

New methods of incarceration: Shenon (1988).

Database structures as containers: Folk and Zoellick (1987).

The idea for containing and retrieving clothing comes from staring at my closet and thinking about what went before. The section on clothing containers could have been far longer. The varieties of clothing storage devices are much greater than presented here. Hangers alone begin to rival the variety of a Darwinian bestiary.

Memory and graded natural categories arose from trying to connect physical and mental storage ideas. The natural categories of Rosch (1975, 1977) were an important concept in putting the ideas together.

For much of the information on firearms and the timed demonstrations, I am indebted to Keith Weaver and Ralph Shields, local experts on antique firearms.

I would also like to thank Valerie Millheim and Jim Buffalo for serving as camera crew during the interview and at the firing range. On discrete, batch, and continuous processing: Weber and Dixon (1989).

17. Procedure's Way

The supermarket essay is very loosely based on some speculations of Perkins (1987) regarding procedural inventions. The last example, a grocery list presented to a "waiter," bears a close similarity to the old call-in order to the neighborhood grocery store. The delivery boy then drops off the food at your apartment. Of course, for ready-to-serve food, a phone-in order system is widely used. Now some restaurants will accept a *faxed order* for food to be picked up— or it can be delivered. Some pizza places will deliver, along with the pizza, a video movie. However, the customer must return the movie the next day. Alternatively, the return can be avoided and the pizza establishment will pick it up the next day, if you place another order. Inventing procedures may be the most common form of invention that most people engage in.

The transition from handcraft to assembly line is based on Weber and Dixon (1989). The written material for the assembly line comes from Giedion (1948) and Ford (1925). The personal view of the assembly line came from my father, Clarence Weber, who worked in the Detroit auto industry during the 1920s until the early 1940s. The importance of switching time is from Weber, Blagowsky, and Mankin (1982) and Weber, Holmes, Gowdy, and Brown (1987).

Regarding the idea of maintenance to extend the life of an artifact, I can find no evidence that tool use by chimps employs anything like maintenance; it seems to be manufacturing and use on the spot (Goodall, 1986). Microwear analysis and maintenance of Stone Age tools is discussed by Semenov (1964). The idea of repairing from afar is discussed in Feder (1991).

Some afterthoughts on maintenance. Some sweaters come with a bit of spare color-matched yarn, but it is usually not attached to the sweater. Christmas tree bulbs often come with extra flashers. My car came with a little bottle of touch-up paint. I have been told that some fancy cars have computers that will automatically tell you when to take your car in for maintenance.

18. Transgenic Myth to Transgenic Mouse

On the first transgenic organism produced by human gene tinkering, a bacterium with a toad gene, see Cohen and Boyer (1988) for a patent based on their scientific work in the 1970s.

Figure 18.1, the mouse with a human oncogene and a large breast cancer tumor, is from Marx (1984). The transgenic mouse with the breast tumor from a human cancer gene is the subject of the first-ever patent for higher life forms, granted to Leder and Stewart (1988). I was unable to obtain a photograph from Leder, so Figure 18.1 is based on the account of Marx.

Figure 18.2, the Sorcerer, is from L. Leakey (1954). The description of the Sorcerer is from L. Leakey (1954, p. 151).

For the rendering of Figures 18.3, 18.4, and 18.5, I want to acknowledge the work of my students S. Palmer and S. Manley.

For domestication of the dog and livestock: Zeuner (1954).

For domestication of the citrus killer-ant and the silkwork: Needham (1971), Temple (1986).

For the taming of bacteria in the genetics engineering lab, see Chakrabarty (1981), the first-ever patent on microorganisms.

For the domestication of the rabies virus by Pasteur, see Dubos (1960), Kolata (1988), Vaughn (1988).

When I talk of evolution as a tree-structure, I am talking about the development of more complex animals. As Margulis and Sagan (1986) point out, a lot of networklike mergings occur at the microbe level. That makes the microbial level more like N^2 in possible pairings, and very similar to the invention of artifacts—where almost anything can in principle combine with anything else. Two differences between microbe and artifact invention remain, however. First, microbial evolution is still likely to be slower than artifactual change because purpose focuses and guides artifacts but not microbes. Second, the considerable speed that microbial evolution does have is due to billions of self-replicating experiments taking place more or less simultaneously. Human invention is a much more serial process with far fewer tries, probably rooted in the serial nature of human consciousness. The study of genetic algorithms (Goldberg, 1989), a new way of thinking about computer programming, attempts to program in a way analogous to bacterial evolution. This looks like an exciting development to me. Human mindware takes on the characteristics of bacterial evolution.

Epilogue: Invention Through the Looking Glass

My discussion of playing on the edge of one's ability is rooted in Perkins (1984) who also discusses the need for wild ideas. For other key ideas I am indebted to Perkins (1992) and Gardner (1991). For the development of clinical ultrasound: Wild (1992) and some personal conversations at the Inventing Mind Conference.

Teaching about technology in the schools is a new and long overdue idea; by this, I mean principles and concepts rather than hardware glitz. Some seminal thoughts on the subject are provided by Johnson (1989) in an American Association for the Advancement of Science curriculum project. The report makes for good reading and points to our society's lamentable lack of educating its citizens on the nature of technology and invention. It does an excellent job of pointing out social aspects of technology. Its one weakness resides in not having a detailed view of invention principles. The comparable AAAS report for mathematics is Blackwell and Henkin (1991), a work highly compatible with the views of invention expressed here.

For the story of Alice, see, of course, Carroll (1960; original 1896).

References

Alexander, C. (1964). *Notes on the synthesis of form.* Cambridge, Mass.: Harvard University Press.

Amato, I. (1990). Diamond fever. *Science News, 138,* 72–74.

American Bison Association. (undated). Nature's bountiful commissary for the Plains Indians. P. O. Box 16660, Denver, Colo. 80216.

Basalla, G. (1988). *The evolution of technology.* Cambridge: Cambridge University Press.

Bijker, W., Hughes, T. P., & Pinch, T. (1987). *The social construction of technological systems: New directions in the sociology and history of technology.* Cambridge, Mass.: The MIT Press.

Blackwell, D. & Helkin, L. (1991). *Mathematics: Report of the Project 2061 Phase I Mathematics Panel.* Washington, D.C.: American Association for the Advancement of Science.

Bordes, F. (1968). *The old Stone Age.* New York: McGraw-Hill.

Campbell, D. T. (1960). Blind variation and selective retention in creative thought as in other knowledge processes. *Psychological Review, 67,* 380–400.

Campbell, J. (1988). *The power of myth.* New York: Doubleday.

Carelman, J. (1984). *Catalog d' objets introuvables.* Paris: Andre' Balland. (First edition, 1969).

Carroll, L. (1960). *Alice's adventures in wonderland & Through the looking-glass.* New York: The New American Library of World Literature. (*Through the looking-glass* originally published 1896).

Chakrabarty, A. M. (1981). Microorganisms having multiple compatible degradative energy-generating plasmids and preparation thereof. U.S. Patent Office #4,259,444. Originally applied for June 7, 1972. For the court case that lead to the patent, see: *Diamond* v. *Chakrabarty*, 447 U.S. 303; 206 U.S.P.Q. 193 (1980).

Childe, V. G. (1954). Wheeled vehicles. In C. Singer, E. J. Holmyard, & A. R. Hall (Eds.), *A history of technology. Vol. 1: From early times to the fall of the ancient empires.* New York: Oxford University Press, pp. 716–729.

Cohen, S. N., & Boyer, H. W. (1988). Biologically functional molecular chimeras. U.S. Patent Office #4,740,470.

Cole, S. M. (1954). Land transport without wheels, roads and bridges. In C. Singer, E. J. Holmyard, & A. R. Hall (Eds.), *A history of technology. Vol. 1: From early times to the fall of the ancient empires.* New York: Oxford University Press, pp. 704–715.

Conot, R. (1979). *Thomas A. Edison: A streak of luck.* New York: Da Capo Press.

Crouch, T. D. (1981). *A dream of wings: Americans and the airplane, 1875–1905.* New York: W. W. Norton.

Crouch, T. D. (1989). *The bishop's boys: A life of Wilbur and Orville Wright.* New York: W. W. Norton.

Crouch, T. D. (1992). Why Wilbur and Orville? Some thoughts on the Wright brothers and the process of invention. In R. J. Weber & D. N. Perkins (Eds.), *Inventive minds. Creativity in technology.* New York: Oxford University Press.

de Mestral, G. (1955). Velvet type fabric and method of producing same. U.S. Patent Office #2,717,437.

de Mestral, G. (1961). Separable fastening device. U.S. Patent Office #3,009,235.

Dennell, R. (1986). Needles and spear-throwers. *Natural History,* October.

Derry, T. K., & Williams, T. I. (1961). *A short history of technology.* New York: Oxford University Press.

Drachmann, A. G. (1967). The classical civilizations. In M. Kranzberg and C. W. Pursell, Jr., *Technology in western civilization. Vol. 1: The emergence of modern industrial society* from *earliest times to 1900.* New York: Oxford University Press.

Drower, M. S. (1954). Water supply, irrigation, and agriculture. In C. Singer, E. J. Holmyard, & A. R. Hall (Eds.), *A history of technology. Vol. 2: From early times to the fall of the ancient empires.* New York: Oxford University Press, pp. 520–557.

Dubos, R. (1960). *Pasteur and modern science.* Garden City, N.Y.: Anchor Books, Doubleday.

Edison, T. A. (1878). Phonograph or speaking machine. U.S. Patent Office #200,521.

Esper, T. (1965). The replacement of the longbow by firearms in the English army. *Technology and Culture, 6,* 382–392.

Fabre, M. (1963). *A history of land transportation* (Vol. 7). New York: Hawthorne Books.

Feder, B. J. (1991). Repairing machinery from afar. *The New York Times,* January 30, p. C7.

Folk, M. J., & Zoellick, B. (1987). *File structures: A conceptual toolkit.* Reading, Mass.: Addison-Wesley.

Forbes, R. J. (1956). Hydraulic engineering and sanitation. In Singer, C., E. J. Holmyard, & A. R. Hall (Eds.), *A history of technology, Vol. 2.* New York: Oxford University Press.

Ford, H. (1925). *My life and work*. New York: Doubleday.

French, M. J. (1988). *Invention and evolution: Design in nature and engineering*. Cambridge: Cambridge University Press.

Friedel, R. & Israel, P. (1987). *Edison's electric light: Biography of an invention*. New Brunswick: Rutgers University Press.

Gardner, H. (1983). *Frames of mind*. New York: Basic Books.

Gardner, H. (1991). *The Unschooled Mind: How Children Think and How Schools Should Teach*. New York: Basic Books.

Giedion, S. (1948). *Mechanization takes command*. New York: Oxford University Press. (Reprinted in 1969 by Norton, New York.)

Gille, B. (1986). *The history of techniques, Vol 1: Techniques and civilizations*. Montreux, Switzerland: Gordon and Breach Science Publishers. (Originally published in French as *Histoire des techniques l'encyclope'die de la ple'iade, c Editions Gallimard. 1978.*)

Gille, B. (1986). *The history of techniques, Vol 2: Techniques and sciences*. Montreux, Switzerland: Gordon and Breach Science Publishers. (Originally published in French as *Histoire des techniques l'encyclope'die de la ple'iade, c Editions Gallimard, 1978.*)

Goldberg, D. E. (1989). *Genetic algorithms in search, optimization, and machine learning*. New York: Addison-Wesley.

Goodall, J. (1986). *The chimpanzees of Gombe: Patterns of behavior*. Cambridge, Mass.: Belknap Press of Harvard University Press.

Gorman, M., & Carlson, B. (1989). Bell, Edison, and the telephone: Invention as a cognitive process. The Inventing Mind Conference, November 1989, Tulsa, OK. To appear in R. J. Weber & D. N. Perkins (Eds.), *Inventive minds: Creativity in technology*. New York: Oxford University Press.

Graf, G. (1987). Diamonds find new settings. *High Technology*, April, pp. 44–47.

Hadamard, J. (1954) *The psychology of invention in the mathematical field*. New York: Dover.

Hofstadter, D. R. (1985). Variations on a theme as the crux of creativity. In D. R. Hofstadter, *Metamagical themas*. New York: Basic Books.

Hounshell, D., & Smith, Jr., J. K. (1988). *Science and corporate strategy*. Cambridge: Cambridge University Press.

Invent America Newsletter. United States Patent Model Foundation, February, 1987

Jehl, F. (1990). *Menlo Park reminiscences: Vol 1. Henry Ford Museum & Greenfield Village*. Printed by Dover Publications (Originally published Dearborn, Mich.: The Edison Institute, c. 1937.)

Johnson, J. R. (1989). *Technology: Report of the Project 2061 Phase I Technology Panel*. Washington, D.C.: American Association for the Advancement of Science.

Johnson, M. (1987). *The body in the mind: The bodily basis of meaning, imagination, and reason*. Chicago: The University of Chicago Press.

Kim, S. H. (1990). *Essence of Creativity: A guide to tackling difficult problems*. New York: Oxford University Press.

Koestler, A. (1964). *The act of creation*. New York: Dell.

Kolata, G. (1987). New Technique May Create Decoy to Halt Spread of AIDS. *The New York Times*, December 18, pp. 1, 14.

Kranzberg, M. & Purcell, Jr., C. (Eds.) (1967). *Technology in western civilization,*

Vol. 1: The emergence of modern industrial society from earliest times to 1900. New York: Oxford University Press.

Lakoff, G. (1987). *Women, fire, and dangerous things: What categories reveal about the mind.* Chicago: The University of Chicago Press.

Langley, P., Simon, H. A., Bradshaw, G. L., & Zytkow, J. M. (1987). *Scientific discovery: Computational explorations of the creative process.* Cambridge, Mass.: The MIT Press.

Leakey, L. S. B. (1954). Graphic and plastic arts. In C. Singer, E. J. Holmyard, & A. R. Hall (Eds.), *A history of technology, Vol. 1*, pp. 144–153. New York: Oxford University Press.

Leakey, M. D. (1971). *Olduvai Gorge, Vol 3.* Cambridge: Cambridge University Press.

Leder, Philip & Stewart, T. A. (1988). Transgenic non-human mammals. U.S. Patent Office #4,736,866. Originally applied for June 22, 1984.

Lenat, D. B. (1978). The ubiquity of discovery. *Artificial Intelligence, 9,* 257–285.

Lenat, D. B. (1983). Toward a theory of heuristics. In R. Groner, M. Groner, & W. F. Bischof (Eds.), *Methods of heuristics* (pp. 351–404). Hillsdale, N.J.: Lawrence Erlbaum Associates.

Littauer, M. A., & Crouwel, J. H. (1979). *Wheeled vehicles and ridden animals in the ancient near east.* Leiden/Kohl: E. J. Brill.

Lloyd, S. (1954). Building in brick and stone. In C. Singer, E. J. Holmyard, & A. R. Hall (Eds.), *A history of technology, Vol. 1.* New York: Oxford University Press.

Macaulay, D. (1988). *The way things work.* Boston: Houghton Mifflin.

MacCurdy, E. (1958). *The notebooks of Leonardo da Vinci.* New York: George Braziller (Originally published by Reynal & Hitchcock, 1939). Quote p. 500; from B 83v.

Margulis, L., & Sagan, D. (1986). *Microcosmos: Four billion years of microbial evolution.* New York: Touchstone.

Marx, J. L. (1984). Tumor-prone mice—and myc. *Science, 226,* p. 823.

McFarland, M. W. (Ed.) 1953). *The papers of Wilbur and Orville Wright. Vol I: 1899–1905; Vol. 2. 1906–1948.* New York. McGraw-Hill.

Miller, D. C. (1934). *The science of musical sounds.* New York: Macmillan.

Minsky, M. (1975). A framework for representing knowledge. In P. H. Winston (Ed.), *The psychology of computer vision.* New York: McGraw-Hill.

Mokyr, J. (1990). *The lever of riches: Technological creativity and economic progress.* New York: Oxford University Press.

Morgan, P. W. (1989). Discovery and invention in polymer chemistry. In R. J. Weber & D. N. Perkins (Eds.), *Inventive minds: Creativity in technology.* New York: Oxford University Press.

Nayak, P. R., & Ketteringham, J. M. (1986). *Breakthroughs!* New York: Rawson Associates.

Needham, J. (with the collaboration of Ling, W.) (1965). *Science and civilzation in China. Vol. 4: Physics and physical technology. Part II: Mechanical engineering.* Cambridge: Cambridge University Press.

Needham, J. (with the collaboration of Ling, W., and Gwei-Djen, L.) (1971). *Science and civilization in China. Vol. 4: Physics and physical technology. Part III: Civil engineering and nautics.* Cambridge: Cambridge University Press.

Needham, J. (with the collaboration of Gwei-Djen, L., and Hsing-Tsung, H.) (1971). *Science and civilization in China. Vol. 6: Biology and biological technology. Part I: Botany.* Cambridge: Cambridge University Press.

Neuburger, A. (1969). *The technical arts and sciences of the ancient.* (Trans. H. L. Brose.) New York: Barnes & Noble. (Originally published in German as *Die Technik des Altertums,* 1930.)

Newell, A., & Simon, H. A. (1972). *Human problem solving.* Englewood Cliffs, N. J.: Prentice-Hall.

Norman, D. A. (1988). *The psychology of everyday things.* New York: Basic Books.

Osborn, A. F. (1963). *Applied imagination* (3rd ed.). New York: Scribner & Sons.

Panati, C. (1987). *Extraordinary origins of everyday things.* New York: Harper & Row.

Pedretti, C. (1987). Introduction. In *Leonardo da Vinci: Engineer and architect.* Quebec: The Montreal Museum of Fine Arts. (Leonardo's helicopter, p. 1. From Paris MS. B, f. 83 v; detail.)

Perkins, D. N. (1981). *The mind's best work.* Cambridge, Mass.: Harvard University Press.

Perkins, D. N. (1984). Creativity by design. *Educational Leadership,* September 18–25.

Perkins, D. N. (1987). *Knowledge as design.* Hillsdale, N.J.: Lawrence Erlbaum Associates.

Perkins, D. N. (1992). *Smart schools: From educating memories to educating minds.* New York: The Free Press.

Perkins, D. N., & Laserna, C. (1986). *Inventive thinking: Teacher manual.* Watertown, Mass.: Mastery Education Corporation.

Perkins, D. N., & Martin, F. (1986). Fragile knowledge and neglected strategies in novice programmers. In E. Subway & S. Lyengar (Eds.), *Empirical studies of programmers,* pp. 213–229. Norwood, N.J.: Ablex.

Perkins, D. N. & Weber, R. J. (1992). Effable invention. In R. J. Weber & D. N. Perkins, *Inventive minds: Creativity in technology.* New York: Oxford University Press.

Petroski, H. (1989). *The pencil.* New York: Knopf.

Philip, C. O. (1985). *Robert Fulton: A biography.* New York: Franklin Watts.

Piggott, S. (1983). *The earliest wheel transport.* Ithaca, N.Y.: Cornell University Press.

Pitt-Rivers, A. L-F. (1906). *The evolution of culture and other essays.* Oxford: Oxford University Press.

Planck, M. (1949). *Scientific autobiography.* New York: Philosophical Library.

Polya, G. (1945, 1957). *How to solve it: A new aspect of mathematical method.* Princeton, N.J.: Princeton University Press.

Polya, G. (1954). *Mathematics and plausible reasoning* (2 vols). Princeton, N.J.: Princeton University Press.

Rabinow, J. (1990). *Inventing for fun and profit.* San Francisco: The San Francisco Press.

Rosch, E. (1975). Cognitive representations of semantic categories. *Journal of Experimental Psychology: General, 104,* 192–223.

Rosch, E. (1977). Human categorization. In N. Warren (Ed.), *Advances in Cross-Cultural Psychology, Vol. 1.* London: Academic Press.

Schneider, K. (1988, April 13). Patent for mouse issued to Harvard. *The New York Times,* pp. 1, 13.

Scott, A. F. (1984). The invention of the balloon and the birth of modern chemistry. *Scientific American, 250,* 122–133.

Semenov, S. A. (1964). *Prehistoric technology.* (Translated by M. W. Thompson). New York: Barnes & Noble.

Shank, R. C., & Abelson, R. P. (1977). *Scripts, plans, goals, and understanding.* Hillsdale: N.J.: Lawrence Erlbaum Associates.

Shenon, P. (1988, February 10). Sensors are used to monitor parolees. *New York Times,* p. A17.

Shlesinger, B. E., Jr. (1987). *How to invent: A text for teachers and students.* New York: IFI/Plenum. (Revised edition of *The art of successful inventing.* 1973.)

Simon, H. A. (1981). *The sciences of the artificial* (2nd ed.). Cambridge, Mass.: The MIT Press.

Simonton, D. K. (1984). *Genius, creativity, and leadership.* Cambridge, Mass.: Harvard University Press

Simonton, D. K. (1988). *Scientific genius.* Cambridge: Cambridge University Press.

Singer, C., E. J. Holmyard, & A. R. Hall (Eds.) (1954–1958). *A history of tech nology,* Vols. 1–5. New York: Oxford University Press.

Smalley, R. E. (1991). Great balls of carbon: The story of Buckminsterfullerene. *The Sciences,* March-April, pp. 22–28.

Stordcur-Yedid, D. (1979). *Les aiguilles a chas au Paleolithique.* Paris: XIIIe supplement to *Gallia Prehistoire.*

Suh, N. P. (1990). *The principles of design.* New York: Oxford University Press.

Temple, R. (1986). *The genius of China.* New York: Simon & Schuster.

Trotter, D. M., Jr. (1991). Photochromic and photosensitive glass. *Scientific American, 264,* 124–129.

Tweney, R. D. (1992). Inventing the field: Michael Faraday and the creative "engineering" of electromagnetic field theory. In R. J. Weber & D. N. Perkins (Eds.), *Inventive minds: Creativity in technology.* New York: Oxford University Press.

U.S. Department of Commerce (1985). *General information concerning patents.* Washington, D.C.: U.S. Government Printing Office.

Usher, A. P. (1954). *History of mechanical inventions.* Cambridge, Mass.: Harvard University Press.

Vaughn, C. (1988, Dec. 3, p. 358). AIDS virus accepts toxic Trojan horse. *Science News, 134.*

Velcro USA, Inc. (Undated). Velcro brand fasteners. Company brochure.

Wallace, D. B., & Gruber, H. E. (Eds.). (1989). *Creative people at work.* New York: Oxford University Press.

Weber R. J. (1992). Stone Age knife to Swiss Army knife: A prototype of invention. In R. J. Weber & D. N. Perkins, *Inventive minds: Creativity in technology.* New York: Oxford University Press.

Weber, R. J., Blagowsky, J., & Mankin, R. (1982. Switching time between overt and covert speech: generative attention. *Memory and Cognition, 10,* 546–553.

Weber, R. J., & Dixon, S. (1989). Invention and gain analysis. *Cognitive Psychology, 21,* 283–302.

Weber, R. J., Dixon, S., & Llorente, A. (in press). Studying invention: The hand tool as a model system. *Science, technology, and human values.*

Weber, R. J., Holmes, M., Gowdy, R., & Brown, S. (1987). Switching between musical response modes: Evidence for global parameter settings. *Music Perception, 4,* 361–372.

Weber, R. J. & Llorente, A. (in press). Natural paths to invention: Reinventing the wheel. *Current Psychology.*

Weber, R. J., Moder, C. L., & Solie, J. B. (1990). Invention heuristics and mental processes underlying the development of a patent for the application herbicides. *New Ideas in Psychology, 8,* 321–336.

Weber, R. J., & Perkins, D. N. (1989). How to invent artifacts and ideas. *New Ideas in Psychology, 7,* 49–72.

Weber, R. J., & Perkins, D. N. (Eds.) (1992a). *Inventive minds: Creativity in technology.* New York: Oxford University Press.

Weber, R. J., & Perkins, D. N. (1992b). The unphilosopher's stone. In R. J. Weber & D. N. Perkins, *Inventive minds: Creativity in technology.* New York Oxford University Press.

Weekly Reader. 1986 award winners. News release, undated.

Weekly Reader. 1987 award winners. News release, undated.

Weisberg, R. (1986). *Creativity: Genius and other myths.* New York: Freeman.

Wentorf, R., Jr. (1992). The synthesis of diamond. In R. J. Weber & D. N. Perkins (Eds.), *Inventive minds: Creativity in technology.* New York: Oxford University Press.

Westrum, R. (1991). *Technologies & society: The shaping of people and things.* Belmont, Calif.: Wadsworth.

White, R. (1986). *Dark caves, bright visions: Life in Ice Age Europe.* New York: American Museum of Natural History.

White, Jr., L. (1962). *Medieval technology and social change.* Oxford: Oxford University Press.

Whitehead, A. N. (1948). *An introduction to mathematics.* New York: Oxford University Press.

Wild, J. J. (1992). The origin of soft tissue ultrasonic echoing and early instrumental application to clinical medicine. In R. J. Weber & D. N. Perkins (Eds.), *Inventive minds: Creativity in technology.* New York: Oxford University Press.

Wile, R. R. (1982). *Journal of the Association for Recorded Sound Collections, 14,* 5–28.

Winner, E. (1984). *Invented worlds: The psychology of the arts.* Cambridge, Mass.: Harvard University Press.

Wundt, W. (1900–1920). *Folk psychology.* Leipzig: Engelmann.

Zeuner, F. E. (1954a). The domestication of plants. In C. Singer, E. J. Holmyard, & A. R. Hall (Eds.), *A history of technology, Vol. 1.* New York: Oxford University Press.

Zeuner, F. E. (1954b) The domestication of animals. In C. Singer, E. J. Holmyard, & A. R. Hall (Eds.), *A history of technology, Vol. 1.* New York: Oxford University Press.

Zwicky, F. (1969). *Discovery, invention, research: Through the morphological approach.* New York: Macmillan.

Index